THE POLITICAL ECONOMY OF INDUSTRIAL PROMOTION

THE POLITICAL ECONOMY OF INDUSTRIAL PROMOTION

Indian, Brazilian, and Korean Electronics in Comparative Perspective 1969–1994

Eswaran Sridharan

Westport, Connecticut
London

Library of Congress Cataloging-in-Publication Data

Sridharan, Eswaran.
The political economy of industrial promotion : Indian, Brazilian, and Korean electronics in comparative perspective 1969–1994 / Eswaran Sridharan
p. cm.
Includes bibliographical references and index.
ISBN 0–275–95418–8 (alk. paper)
1. Electronic industries—Government policy—India. 2. Electronic industries—Government policy—Brazil. 3. Electronic industries—Government policy—Korea (South). 4. Industrial promotion—India. 5. Industrial promotion—Brazil. 6. Industrial promotion—Korea (South). I. Title.
HD9696.A3I753 1996
388.4'7621381'095—dc20 95–43766

British Library Cataloguing in Publication Data is available.

Library of Congress Catalog Card Number: 95–43766
ISBN: 0–275–95418–8

First published in 1996

Praeger Publishers, 88 Post Road West, Westport, CT 06881
An imprint of Greenwood Publishing Group, Inc.

Printed in the United States of America

The paper used in this book complies with the Permanent Paper Standard issued by the National Information Standards Organization (Z39.48–1984).

10 9 8 7 6 5 4 3 2 1

TO
DIVYA
AND MY PARENTS

Contents

Tables

Preface and Acknowledgments

This book is an interpretative comparison of the political economy of promoting a new industry—electronics—in three countries, India, Brazil, and Korea, over a period of a quarter of a century from 1969 to 1994. Electronics is broadly defined as the entire microelectronics-based complex of industries comprising consumer electronics, computers and software, control instrumentation and industrial electronics, communications and broadcasting, defense/aerospace electronics, and electronic components. Technologically, these are all increasingly convergent today. India, Brazil, and South Korea are, industrially and technologically, three of the most advanced countries in the developing world. All three countries systematically began to promote the development of local electronics industries at the beginning of the period concerned, each with its own initial conditions, its political economy of development, and its distinctive strategy. The early 1990s were a period of transition for all three in electronics policy and development. Korea is the archetype of successful export-led growth, while Brazil and India are both large, inward-looking economies but with quite distinct brands of import-substituting industrialization and technology development. Our primary focus is on India, where we did our fieldwork, with Korea and Brazil as comparative reference points for which we rely on the vast literature and data sources available.

This book's main strength is intended to be its synoptic overview of three-country experience, focusing not only on the technicalities of the electronics case but situating it in the context of a broader debate in the comparative political economy of development on the conditions and modes of successful and unsuccessful state intervention in industrialization. We focus specifically on the debate around the concept of the strategic capacity of developing-country states to effect industrial transformation. As a three-country, quarter-century interpretative comparison of the political economy of industrialization that defines electronics broadly, the book covers a vast canvas. Consequently, and because of the constraints of space, drastic compression of the original material was required. The book, therefore, focuses on the "big story," referring to all too many smaller stories that are part of the former only selectively and in brief. In particular, the telecommunications industry—a vast industry in itself, policy for which is made in India and most countries by a separate government department—is treated only insofar as it affects the development of the rest of the electronics industry.

This book was written in its present form while I was on the faculty of the Centre for Policy Research (CPR), New Delhi. It originated as a doctoral dissertation at the Department of Political Science at the University of Pennsylvania, completed in 1989. Fieldwork in India was undertaken in 1986–87 and in 1988. While the book was being updated and thoroughly rewritten at the CPR alongside other full-time commitments, further fieldwork was sporadically undertaken during 1991–95. I am indebted to several institutions for their financial and institutional support. The Department of Political Science, University of Pennsylvania, supported me with fellowships and teaching and research assistantships during my doctoral program and awarded me the Landy fellowship for dissertation fieldwork in 1986–87. The CPR facilitated the writing of the book in its present form. The Institute for Developing Economies, Tokyo, granted me a Visiting Research Fellowship during the first quarter of 1993, where I wrote a shorter version of the present book, which I completed for submission for publication while a Visiting Fellow at the Centre for International Studies, London School of Economics, in the last quarter of 1994, while there on another project.

I am also deeply indebted to a large number of individuals over the years. Francine R. Frankel, Youssef Cohen, and Anil Deolalikar supervised my doctoral dissertation. Joanne Gowa read part of my dissertation draft and gave me detailed and stimulating comments on theoretical issues. P. V. Indiresan gave me detailed comments, on a shorter version of this book, that helped correct some errors. In India during my fieldwork I received the help and cooperation of over fifty persons in industry, government, and industry associations around the country who granted me interviews, including several former and then-serving Secretaries and other then-incumbent and former officials of the Department of Electronics and Electronics Commission and public sector enterprises. I am deeply grateful to all of them, especially those of the latter organizations who granted me interviews on condition of anonymity. I am therefore unable to make attributions by name to those to whom I am most indebted. However, I take the liberty of mentioning the names of Ashok Parthasarathi, Sam Pitroda, M. G. K. Menon, A. S. Rao, C. R. Subramanian, S. R. Vijayakar, P. R. Dastidar, N. Seshagiri, N. Vittal, and Pronab Sen, among those who granted me detailed, in many cases, repeated interviews.

I am indebted to Peter Evans for stimulating discussions and useful material as well as for doing me the honor of sending me the manuscript of his book for my comments. It was too late for me to engage with his book's argument, although I have cited his earlier writings extensively. For invaluable material which saved me a lot of time I am grateful to G. Balachandran, Hans-Peter Brunner, John Echeverri-Gent, Hideki Esho, Claudio Frischtak, Geoff Gowen, and Michiro Yatani. Ravi Chopra and Ashutosh Varshney helped in various ways. I am, of course, solely responsible for all facts, figures, and interpretations in this book.

Last but not least, I owe Divya, and my parents and parents-in-law, far more than the usual platitudes about their love, understanding, and patience.

Abbreviations and Acronyms

ASIC	application-specific integrated circuit
b&w	black-and-white
BARC	Bhabha Atomic Research Centre
BEL	Bharat Electronics Limited
BHEL	Bharat Heavy Electricals Limited
BICP	Bureau of Industrial Costs and Prices
BNDE	National Economic Development Bank
BNDE(S)	National Economic (and Social) Development Bank
BNH	National Housing Bank
CAD	computer-aided design
CAPES	Campaign for Improvement of High Manpower
CAPRE	Commission for Coordination of Electronic Processing Activities
CDAC	Centre for Development of Advanced Computing
CDMA	code division multiple access
CDOME	Council for the Development of Materials
CDOT	Centre for Development of Telematics
CEERI	Central Electronics Engineering Research Institute
CETUC	Center for Telecom Studies
CIIE	control instrumentation and industrial electronics
CMC	Computer Maintenance Corporation
CMOS	complementary metal-oxide semiconductor
CNPq	National Center for Research and Development

COBRA	Computadores Brasileira
CONIN	National Council on Informatics and Automation
CPCT	Centro de Estudos em Politica Cientifica e Tecnologica
CPqD	Center for Research and Development
CPU	central processing unit
CSIO	Central Scientific Instruments Organization
CSN	National Security Council
CTB	Companhia Telefonica Brasileira
CTI	Centre for Informatics Technology
DACOM	Data Communications Corporation of Korea
DAE	Department of Atomic Energy
DDC	distributed digital control
DEPIN	Department of Informatics and Automation Policy
DOE	Department of Electronics
DOT	Department of Telecommunications
DRAM	dynamic random access memory
DRDO	Defence Research and Development Organization
DTA	Domestic Tariff Area
EC	Electronics Commission
ECIL	Electronics Corporation of India Limited
EDP	electronic data processing
EEPROM	electrically erasable programmable read-only memory
EIAK	Electronic Industries Association of Korea
EIP	Electronics Information and Planning
ELCINA	Electronic Component Industries Association
EMBRATEL	Brazilian Telecommunications Company
EOU	export-oriented unit
EPABX	electronic private automatic branch exchange
EPB	Economic Planning Board
EPROM	erasable programmable read only memories
EPZ	export processing zone
ERDC	Electronics Research and Development Center
ERNET	Education and Research Network
ESS	electronic switching system
ET&T	Electronics Trade and Technology Development Corporation
ETRI	Electronics and Telecommunications Research Institute
FDI	foreign direct investment
FDM	frequency division multiplexing
FDS	frequency division switching
FERA	Foreign Exchange Regulation Act

FINAME	Special Agency For Industrial Finance
FNDCT	National Fund for Scientific and Technological Development
FTZ	free-trade zone
FUNTEC	Special Fund for Technology
GaAs	gallium arsenide
GDP	gross domestic product
GEICOM	Inter-Ministerial Executive Group on Components and Materials for Communications
GTC	General Trading Company [*sogo shosha* (Japan)]
HAL	Hindustan Aeronautics Limited
HMT	Hindustan Machine Tools
IBICT	Brazilian Institute of Information Science and Technology
IC	integrated circuit
ICIM	International Computers Indian Manufacture
IISc	Indian Institute of Science
IITs	Indian Institutes of Technology
IMF	International Monetary Fund
IMSC	Inter-Ministerial Standing Committee
INPI	National Institute of Industrial Property
IPAG	Information, Planning and Analysis Group
ISI	Import-substitution industrialization
ITI	Indian Telephone Industries
ITMA	Indian Television Manufacturers Association
KAIST	Korea Advanced Institute of Science and Technology
Kbps	Kilobits/second
KETRI	Korea Electrotechnology and Telecommunications Research Institute
KIET	Korea Institute of Electronics Technology
KTA	Korean Telecommunications Authority
LAN	local area network
LDC	less developed country
LED	Laboratory for Electronic Devices
LME	Laboratory for Microelectronics
LRDE	Electronics and Radar Development Establishment
LSI	large-scale integration
MAIT	Manufacturers Association for Information Technology
MAX	main automatic exchange
MCT	Ministry of Science and Technology
Minicom	Ministry of Communications
MIPS	million instructions/second
MITI	Ministry of International Trade and Industry
MNC	multinational corporation

MOC	Ministry of Communications
MODVAT	modified value-added tax
MOS	metal-oxide semiconductor
MOST	Ministry of Science and Technology
MOTIE	Ministry of Trade, Industry and Energy
MRTP	Monopolies and Restrictive Trade Practices Act
MTI	Ministry of Trade and Industry (later MOTIE)
MTNL	Mahanagar Telephone Nigam Limited
NAIS	National Administrative Information System
NAL	National Aeronautical Laboratory
NASSCOM	National Association of Software and Service Companies
NBER	National Bureau for Economic Research
NCA	National Computerization Agency
NDP	Net Domestic Product
NEC	Nippon Electric Co.
NIC	newly industrializing country
NMC	National Microelectronics Council
NPL	National Physical Laboratory
OECD	Organization for Economic Cooperation and Development
OEM	original equipment manufacturer
P&T	posts and telegraphs
PCB	printed-circuit board
PCM	pulse code modulation
PLANIN	National Plan for Informatics and Automation
PTT	post, telegraph, and telephone
RAM	random access memories
RAX	rural automatic exchange
ROM	read-only memories
SAMEER	Society for Applied Microwave Engineering and Electronics Research
SCL	Semiconductor Complex Limited
SDM	space division multiplexing
SDS	space division switching
SEEPZ	Santa Cruz Electronics Export Processing Zone
SEI	Special Secretariat for Informatics
SEPLAN	Planning Secretariat
SNDCT	National System of Scientific and Technological Development
SNI	National Intelligence Agency
SPC	stored program control
SSI	small-scale industry
SST	Samsung Semiconductor and Telecommunications

TBDF	transborder data flows
TCS	Tata Consultancy Services
TDM	time division multiplexing
TDMA	time division multiple access
TDS	time division switching
TELEBRAS	Telefonicas Brasileiras
TEMA	Telecom Equipment Manufacturers Association
TIFR	Tata Institute for Fundamental Research
TRC	Telecommunication Research Centre
TV	television
UHF	ultra-high frequency
ULSI	ultra-large-scale integration
Unicamp	University of Campinas
VA	value-addition
VCR	video-cassette recorder
VLSI	very-large-scale integration
VSNL	Videsh Sanchar Nigam Limited

THE POLITICAL ECONOMY OF INDUSTRIAL PROMOTION

1

The State of the Debate on Development and the State

THE STATE, STRATEGIC CAPACITY, AND INDUSTRIAL PROMOTION

The Old Debate: Import-Substitution versus Export-Led Growth and Economic Liberalization

Beginning in the 1970s, there was a worldwide swing in intellectual and political fashion toward market-oriented economic policies and away from state interventionism. This resulted from three simultaneous economic and political developments in the so-called First, Second, and Third Worlds. (1) Keynesian macroeconomic management and social–democratic welfare-statism proved ineffective in the stagflationary 1970s. (2) The growth slowdown in socialist economies from the early 1970s reflected the end of the era of "extensive" growth and the crisis of autarkic central planning. (3) The growth performance of developing countries following import-substitution industrialization (ISI) was unimpressive compared to that of those following export-oriented policies in the context of general economic liberalization. These developments gave rise to a debate about the relative efficacy of ISI and export orientation as industrialization strategies. They also raised the related issue of state planning and interventionism vis-à-vis economic liberalism.

The debate revolved around a number of controversial questions. Was ISI really exhausted and export orientation and opening up of the economy necessary? What was the effect on the rate and content of growth? What happened to the growth and commodity composition of exports? What had been the role of foreign direct investment (FDI)? Was such investment necessary for export-oriented growth? What happened to capital goods production and the growth of technological capability? What had been the results for equity and basic needs? Were these "success stories" sustainable, as might be the case in Korea, or were they ultimately doomed to failure? Furthermore, were such policies and "success stories" replicable in other developing countries? Or was there a fallacy of composition involved? Was the transition of the newly industrializing countries (NICs) historically unique?

Lastly, and most importantly, what had been the role of the state in the export-led growth success stories? Was the state's role really that of a noninterventionist liberal state as had been made out by Balassa, Krueger, and Westphal among others, reaping the benefits of specialization in line with its comparative advantage on Heckscher–Ohlin–Samuelson lines?[1] Or have these states been covertly interventionist along mercantilist or corporatist lines to promote competitiveness in domestic firms as argued by theorists of the capitalist developmental state?[2] How have the more liberal NICs fared compared to the more interventionist ones over the longer term, and what have been the political prerequisites and evolving political factors, domestic and international, that explain the extent and nature of state intervention to promote industrial competitiveness? In the globalizing world economy of today, industrial efficiency and competitiveness, have become imperative whatever the overall economic strategy, and whatever the political system. Without such competitiveness any strategy will founder and end up in a debt crisis. Indeed, it should have been clear even in the 1960s and 1970s that efficient ISI is the only viable ISI.

The New Debate: The Political Economy of Industrial Strategy for Competitiveness

The debate on the political economy of economic liberalization thus evolved into one on the political economy of industrial strategy for competitiveness, focusing on explaining the kind of industrial promotion undertaken by the state. Strategies have ranged from almost classic free-market as in Hong Kong or Singapore, to highly state-managed in favor of domestic firms vis-à-vis multinational corporations (MNCs), also described as mercantilist, as in Korea, to the Brazilian triple alliance of state, local, and foreign capital, to classic ISI led by public and domestic private enterprise as in India.[3] The question then becomes: what are the political conditions necessary for, or political constraints on, the "right" or effective state intervention—that is, policy—to develop a competitive industrial capability?

We briefly review in this chapter the earlier debates on ISI, export-led growth, and economic liberalization in Brazil and Korea and summarize their implications in general and for India in particular. We then introduce the concept of the developmental state arising from the changed debate and the experience of East

Asia. In subsequent chapters, we take up the case of the electronics industry in Korea, Brazil, and India for a comparative study as a particularly appropriate example of the dilemmas of industrial policy. This study is essentially an *interpretative comparison* of the political economy of electronics strategy in India in relation to those of Brazil and Korea, based on field research in India and the literature and data available on Korea and Brazil.

Korea and Brazil are important as comparative reference points for India as both enjoyed very high growth rates of output and exports, unlike India, since the crisis of the first round of ISI in the 1960s. Both were touted (Brazil until the early 1980s) as economic miracles, owing their success to export orientation or openness to foreign capital. Korea abandoned ISI for export-led growth, while Brazil's ISI was relatively open to FDI, unlike India's. Like India, both are among the technologically most advanced less developed countries (LDCs). Korea has become the developing world's largest exporter of manufactures, while Brazil has the nonsocialist developing world's largest GNP and largest capital goods production. And even if Korea is said to be an atypical case not comparable to India, Brazil is an intermediate and more comparable case, being large in area and population and having a sizeable domestic market. Korea and Brazil are also an important contrast in that Brazil's boom went bust in the debt crisis following the oil shock of 1979 while Korea rode out the shock, successfully implementing stabilization and structural adjustment policies, and continued to grow rapidly. This contrast indicates crucial differences in the nature of state intervention and economic development.

Our aim is to compare the Indian with the Korean and Brazilian strategies to try to see what the political requirements and constraints of the successful Korean route were and whether, after explaining the evolution of Indian policy, it would have been possible to have followed such a route or the rather different, less state-regulated Brazilian ISI route for faster industrial growth and competitiveness. What kinds of capacities were required for the Korean strategy? Did India have the capacity for a more effective alternative strategy within the constraints and pressures operating?

The Neoclassical Critique of ISI and Advocacy of Export Orientation

The earlier ISI versus export-led growth debate produced a neoclassical critique of ISI, and a positive assessment of the relative merits of export-oriented free-trade policies, arising mainly out of the massive NBER, World Bank, and OECD studies on trade regimes. It can be summarized as follows.[4] ISI created inefficiencies through the misallocation of resources due to tariff protection, domestic entry barriers, and industrial credit subsidies encouraging the use of capital relative to labor, contrary to the comparative advantage of LDCs deriving from their labor-abundant factor endowment. Inappropriate investments in capital-intensive industries and technologies followed, resulting in slow employment growth. Underutilization of capacity often emerged because of the narrowness of the home market for the "high-income characteristic" goods produced. Agriculture was neglected due to industrial protection shifting the terms of trade against it, biasing the direction of investment in favor of industry. A bias against exports also prevailed from the maintenance of a higher exchange rate than would have been the case under a free-trade regime. Combined with the import-intensity of ISI, that is, import of capital and intermediate goods, this led to persistent trade and balance of payments deficits, often inviting IMF-imposed stabilization policies. Growth tended to peter out after a brief boom. This was referred to as the exhaustion of ISI.

The remedy for such an impasse was said to be a strategy of growth in line with comparative advantage along orthodox lines, intensive in relatively abundant factors of production, as would be the case in exports of labor-intensive manufactures. This could be achieved by getting the LDC's distorted factor prices "right" by liberalization of the economy, including removal of price controls, trade barriers, and credit subsidies, and instituting realistic exchange rates and export incentives. This would lead to greater employment, better income distribution, reduced poverty, and improved balance of payments. Export orientation was also expected to attract FDI along labor-intensive lines for assembly and reexport of manufactures, reinforcing the favorable effects of liberal trade policies, while avoiding the immiserizing effects of tariff-jumping, ISI-induced FDI.[5] The strategy followed by the NICs, especially Korea, was recommended to other LDCs. It was held that export-oriented liberal economic policies produced far better results in terms of growth, equity, basic needs, and balance of payments stability than ISI.

Implications for Dependency Theory

Early formulations of dependency theory expressed pessimism about the possibility of Third World countries breaking out of the international division of labor and moving up the ladder of the world industrial hierarchy. The experience of Korea and other East Asian NICs, and to a lesser extent Brazil in the 1970s, seemed to refute their arguments. These developments led dependency theorists to reformulate their ideas to allow for the possibility of some development (but still, in the final analysis, dependent development) without quite accepting the critique of ISI and the public sector by neoclassical advocates of economic liberalization. These writers can be considered "optimistic" *dependencistas*. They attributed the success of East Asian export-led growth and the somewhat lesser success of Brazilian exports and capital goods industrialization not to liberalization but to effective state-led strategy. Peter Evans identifies *state promotion of national industry* as the key transformative factor in overcoming the domination of metropolitan capital: "Central to such a change in the political balance was the growth of a stronger local state apparatus. Even with industrialization under way an active state apparatus was still required to counteract the tendency of international capital to centralize newer, higher-return kinds of industrial activity in the core."[6]

Other scholars, working mainly on East Asia, have gone beyond dependency to a postdependency theory that allows for the possibility of a breakout from underdevelopment and dependency by *strategy-led development*. The strengthening of the bargaining power of the local state and its use in the interests of nationally owned industry is what makes renegotiation of the relationship with metropolitan capital possible. Crucial to such strategy-led development is the possession by peripheral states of what Frederic Deyo, in a neat conceptual innovation, defines as *strategic capacity*.[7]

Strategic Capacity: Extending the Concept

The concept of strategic capacity originated in the debate on the political economy of East Asian industrialization. We approach the concept via a brief overview of the debate. It is now commonplace to argue that the world industrial hierarchy is not rigid but allows room for successful maneuver and upward mobility by the better-placed developing countries, given appropriate state strategy. Postdependency theorists essentially argue that developing countries can, given appropriate state intervention, strike better bargains with MNCs and developed countries and reorganize their economic policy regimes, institutions, and firms to absorb and build upon imported technology to create comparative advantage, becoming technologically more self-reliant in the process, and eventually to compete in hitherto developed world industries. Several NICs have emerged as significant producers and exporters of more advanced manufactures including capital goods, electronics, automobiles, armaments, and chemicals.

What was behind the success of NIC development of new industries, including significant exports of such products? Evans argued early on in the case of Brazil that state promotion of national industry was the key factor. Protection, public investment in industry, infrastructure, and science and technology, and state regulation of and renegotiation with MNCs were the main instruments used. A broadly similar argument can be made for India. In both cases, such creation of new industries not only reduced import-dependence and foreign-firm dominance but led to significant learning effects, technological depth, and human resource development, creating the potential for future competitiveness.

However, the political economy of state intervention in Korea and the other East Asian NICs has come in for more detailed theorizing as scholars attempted to analyze and explain how these states achieved the success that they did. Stephan Haggard and Chung-in Moon emphasized that the Korean type of state intervention to promote the competitiveness of domestic firms depended on two state characteristics—state autonomy and state capacity. State autonomy referred to the autonomy of the state from not only society in general, but even and especially from the dominant socioeconomic groups, including the giant conglomerates. State capacity referred to the state's ability to formulate and implement economic plans, particularly its ability to impose its policy priorities on the dominant groups.

Chalmers Johnson suggested a similar logic of the "capitalist developmental state" linking Korea and Taiwan with Japan, the original model of such a state, which centers on the interaction of the state and the private sector. The state tries to

> manipulate the inputs into the decision-making processes of privately owned and managed enterprises in order to achieve developmental goals, but the content of its inputs is continuously affected by feedback on profit-and-loss conditions, export prospects, raw material costs and tax receipts. The intent of the private system is to maximize profits, limit risks, and achieve stable growth given the political–economical environment in which it must operate, but its decisions on products, markets and investments are continuously affected by changing costs and availability of capital, export incentives, licensing requirements, and all the other things government manipulates.[8]

Johnson saw the "capitalist developmental state" as "market-conforming" rather than "market-repressing" in its methods of intervention. Its key feature was "stable rule by a political–bureaucratic elite *not acceding to political demands that would undermine economic growth*" (emphasis added).[9]

Deyo goes further to introduce the concept of "strategic capacity" of the state. He defines strategic capacity as consisting of two elements—political closure and economic institutional consolidation. These are essentially elaborations and refinements of the concepts of state autonomy and capacity employed by Haggard and Moon. Political closure refers to the exclusionary character of the state. It is characterized at one level by the autonomy of the state from foreign interests as well as from the popular sector, thus narrowing strategy determination to domestic dominant groups. At a second level it is characterized by state autonomy from all groups outside the state apparatus, even the socially dominant groups. The *exclusionary* character of political closure as the condition necessary for strategic capacity suggests, although Deyo does not argue this, a contradiction between strategic capacity and democratic political regimes.

The second precondition is economic institutional consolidation. The most effective institutional configuration is the concentration of both decisional or strategic, and operational or implementation, authority in one or a few powerful state agencies. These perform the functions of mediating external linkages, laying down the terms and conditions of FDI, loans, and technology transfer as well as managing political relations. They also perform the functions of policy implementation and political control on behalf of the chosen strategy. Taken together, political closure and economic institutional consolidation give rise to strategic capacity. The latter is enhanced by the existence of peak private sector organizations that may be used as "strategic levers" for policy implementation. Deyo argues that the capitalist developmental states of East Asia, preeminently Korea and Taiwan, had this strategic capacity.

Robert Wade goes beyond the developmental state construct to develop a theory of the "governed market" to explain the success of Japan, Korea, and Taiwan. The developmental state is not merely "market-conforming" in its intervention but "market-governing." The theory has three levels of causation. At the proximate causal level, economic success was the result of high investment rates and rapid technological progress on lines other than what free market signals would have produced, due to state guidance or "governance" of the market. This was achieved because of (1) high investment rates, fast turnover of machinery, and transfer of new technologies into production; (2) more investment in certain key industries than would have occurred without state support and direction; (3) exposure to global competition due to government support conditioned on exports.

At a second causational level was the set of government policies that governed market processes and directed firms' activities. The third level—the organization of the state and the private sector and their conditioning social–coalitional base and international environment—described and explained the developmental state and its strategic capacity to govern the market. Wade emphasizes the corporatist and authoritarian character of the state and its essential dominance over the private sector.

Wade classifies governance of the market into "leadership" and "followership" of the market, big and small, in sectoral policies, and he stresses its key role in Korean and Taiwanese industrial upward mobility and export success. Leading the market occurs when "the government takes initiatives about what products and technologies should be encouraged, and puts public resources or public influence behind these initiatives."[10] "Big leadership" happens when state initiatives are major enough to alter significantly investment and production patterns in an industry. "Small leadership" happens when state initiatives are too lacking in resources or influence to make a difference. "Big followership" of the market by the state is when state assistance helps firms significantly scale up the investments they have chosen to make, while "small followership" is when the state helps firms do what they would have done anyway.

Alice Amsden essentially concurs with strategy-led development theories about Korea but emphasizes the export conditionality of state subsidization and, along with Ashoka Mody, the importance of the conglomerate form of industrial organization for the facilitation of long-range planning, technological learning, export orientation, risk-taking, and, ultimately, indigenous innovation.[11]

Peter Evans and Paulo Bastos Tigre argue, in what amounts to a refinement of the strategic capacity concept without making any reference to it, that

> a composite of institutional factors is necessary for effective pursuit of technologically ambitious industrial transformation. Support for the policy must have both breadth and coherence within the state apparatus, rather than being encapsulated in particular fiefdoms. The state must be able to bring to bear a range of policy instruments in a focused way; this implies a diversified but coordinated organizational

base. Equally important are relations with the private sector that support risky initiatives and counteract propensities to choose short-run strategies that undercut long-range developmental prospects. Finally, the state must have the capacity to facilitate international alliances that foster rather than constrain indigenous initiatives.[12]

They also stress the importance of a strong domestic private sector. Evans and Tigre thus emphasize coordination of state agencies for policy coherence over economic institutional consolidation in Deyo's sense. We would argue that in certain circumstances, well-coordinated though diversified policy agencies can substitute for the consolidation of economic policy in one or a few key institutions.

Evans, posing a question related to the theme of this chapter, asks, "What role makes sense for the state in high technology industries?"[13] He sees three possible, overlapping roles—regulator, producer, and promoter, or, in his words, custodian, demiurge, and handmaiden to entrepreneurship, respectively. He finds that in the NICs there is a shift during the 1980s from the regulator and producer roles to more promotional or "handmaidenly" approaches and, furthermore, that the latter "enhance the prospects of industrial transformation in high-technology industries."[14]

"Handmaidenly" approaches consist of providing infrastructure, support to R&D in the private sector and in public–private collaboration, human resource development, information support, and so forth that lower the risks to private enterprise of moving into technologically more advanced products, without entirely shielding them from competitive pressures.

In sum, the above theorizing on the political economy of industrial strategy to develop new technology industries in developing countries highlights the key role of *appropriate* state intervention. What is appropriate is inherently variable in content and changing with the circumstances. It could variously mean leading or following the market, handmaidenly support, traditional selective protection, state enterprise, and so forth, depending on the kind of industry, the stage of development, the organization of the state, and domestic and international economic and political factors. And because of the changing requirements of appropriateness, the capacity of the developmental state to *reconstruct* itself, stressed by Evans and Tigre, is crucial.

This brings us to another key dimension of strategic capacity—*policy flexibility*. Following Baldev Raj Nayar we argue that flexibility of policy action is a crucial dimension of the capacity to make and implement industrial policy, and that this is a *path-dependent variable*.

The use of certain policy instruments at one time may condition subsequent state capacity; it may do so not just by precedent-setting accumulation of instruments, but rather by imparting rigidity to the policy-response structure and making the state less effective. Important therefore in any assessment of state capacity is not merely the number and range of policy instruments, but the flexibility the state has in using them. Thus deregulation and non-intervention may also truly represent state capacity.[15]

Nayar adds,

Focusing on economic policy alone, the possible instruments at the command of the state may be placed along a continuum of increasing intensity of state involvement in the economy, starting from (1) monetary and fiscal policy through (2) state regulation of economic activity to (3) state ownership of the means of production . . . it can be tentatively hypothesised that the degree of flexibility in policy response is likely to be inversely proportional to the increasing intensity of state involvement . . . policy flexibility is likely to be inhibited by technical and organizational complexity, which increases as one proceeds along the continuum. More significantly, the use of different economic instruments creates concrete institutional interests within the state, which are likely to seek the perpetuation and expansion of policies from which they benefit . . . how things are done in the past also influences policy outcomes; that is, policy tends to be path-dependent.[16]

Central to our analysis in this chapter of the factors behind states' ability to intervene *appropriately* is the rich concept of "strategic capacity" introduced by Deyo. Our extension of the concept, as above, synthesizes the work of several others so as to refine and broaden it beyond the East Asian cases to accommodate a variety of strategies that can be appropriate at different times and places.

We add two refinements to the strategic capacity concept:

1. We introduce the notion of *strategic priorities* and, derivatively, the grand strategy of the state elite, taking the concept of strategic capacity beyond merely *capacity in a mechanical or administrative sense*; the argument is that states with similar capacities may choose entirely different strategies depending upon the state elite's strategic priorities. The latter and its determinants condition strategic capacity in the former sense and are therefore a vital dimension of the concept.
2. While Nayar sees policy flexibility as *interest-constrained*, we further refine the concept from the Indian electronics case by adding the notion of *ideological constraints*, posed by the ideological legacy and past positions taken by leaderships, to policy flexibility.

The theoretical upshot of the Korean experience of export-led growth not dependent on MNCs when seen in the context of the conceptual frameworks discussed above is that state-organized strategy-led development that breaks out of the bonds of dependency appears possible even without revolutionary change and delinking from the capitalist world economy. This much is conceded even by the "optimistic" *dependencistas* like Evans, though he emphasizes the domestic political prerequisites and the external environment as necessary conditions. We need to be able to understand better these necessary conditions and the possibility of their being met if successful strategy-led development is to be replicable in the semi-industrialized developing world in general.

THE KOREAN "MIRACLE" AND BRAZILIAN ISI IN COMPARISON TO INDIAN ISI: THE EMPIRICAL RECORD

We shall now examine the empirical record of Korea since the switch to export orientation from 1962 on and Brazil's growth since the macroeconomic stabilization of 1964 that set the stage for a new round of ISI, and try to draw some tentative theoretical and policy conclusions about export-led growth and economic liberalization versus ISI.[17] Growth rates of GDP in Brazil were 9.0% over 1965–80 and 2.7% in the 1980s, respectively; and of GNP per capita, 5.1% from 1960 to 1981, slowing to 0.5% over 1980–91. Per capita GNP reached $2,220 in 1984, doubling in real terms since the mid-1960s and was $2,770 by 1992. Industrial production quadrupled in the same period. Manufacturing industry, after growing at 8.7% for 1970–79, crashed to 1.7% over 1980–90 during the prolonged debt crisis. The percentage distribution of the labor force in agriculture/industry/services shifted from 49/20/31 to 31/27/42 over 1965–80, and urbanization grew from 46% to 68% (and to 77% by 1992) in the same period, despite population growth at 2.8%, 2.1%, and 2.0% in the 1960s, 1970s, and 1980–92, respectively. Both shifts indicate the rapidity of industrial and overall economic growth over 1965–90, especially up to 1980.

Brazilian growth after the 1964 coup was led by consumer durables, especially the automotive sector, and after 1967 by government-financed construction. After 1974 there was a tremendous rise in the output of capital goods (sevenfold growth between 1970 and 1979). Export growth was rapid from the mid-1960s, with exports growing from 5% to 9% of GDP over 1960–81 and 10% by 1992. Merchandise exports grew at rates of 5%, 8.7%, and 4.0% in the 1960s, 1970s, and 1980s, respectively. The commodity composition of exports shifted dramatically toward manufactures: the primary/machinery–and–transport–equipment/other manufactures (of which, textiles) breakup rose from 92/2/7(1) to 47/18/35(3) over 1965–90. Production and export of capital goods showed a significant increase, with exports of them rising to 29% of all manufactured exports by 1978. This is significant as the capital goods sector is the main vehicle and indicator of technological progress. Brazil became the largest capital goods exporter in the Third World by 1980, even though its total manufactured exports were only half those of Korea.[18]

South Korea's GDP growth rates were 8.6%, 9.1%, and 9.7% in the 1960s, 1970s, and 1980s, respectively, and 9.9% over 1965–80. Its per capita GDP growth rate was 6.9% over 1960–81 and 8.5% over 1980–92, reaching $1,700 in 1981 (not far behind Brazil) and $6,790 by 1992 (two-and-a-half times Brazil's). Korean industry grew at an astounding 17.2% and 17.4% in the 1960s and 1970s, respectively, its manufacturing sector at 17.6% and 15.6% in the same periods, again higher than Brazil's, including 18.7% over 1965–80. Over 1980–90, manufacturing slowed down to a still amazing 12.7%. The percentage distribution of GDP in agriculture/industry (manufacturing)/services changed dramatically over

1965–80–90, from 48/25(18)/27 to 15/41(30)/44 to 9/45(29)/46. The percentage distribution of the labor force in agriculture/industry/services changed from 55/15/30 to 36/27/37 to 18/35/47 over 1965–80–90. The shifts in both measures were greater than Brazil's. Urbanization shot up from 28% (1961) to 56% (1981) to 74% (1992). This was despite demographic growth rates of 2.6%, 1.7%, and 1.1% in the 1960s, 1970s, and 1980–92, respectively.

Growth was export-led after 1962, especially after 1967. The growth rate of merchandise exports during 1965–80 and during 1980–90 were 27.2% and 12.8%, respectively. The commodity composition of exports shifted dramatically toward manufactures: the primary/machinery–and–transport–equipment/other manufactures (of which, textiles) ratio changed from 40/3/56(27) to 7/37/57(22) over 1965–90, revealing the shift to complex manufactures and away from textiles. Most of the machinery and transport equipment sector consisted of electronics, steel, automobiles, and chemicals by the late 1980s and early 1990s. By 1988 electronics had overtaken textiles as the leading export industry. In absolute terms manufactured exports were $15.722 billion in 1980 compared to Brazil's $7.77 billion, and $71 billion (out of total exports of $76.4 billion) in 1992 compared to Brazil's $20 billion (out of total exports of $36 billion).

Capital goods production rose almost fifteenfold over 1970–79, especially after 1973. One third of capital goods produced were exported in 1979; unlike Brazil, two-thirds were exported to developed countries and included both electrical and nonelectrical machinery, transport equipment, and electronic goods. This gives an indication of the extent of technological progress.

India's growth record was modest in comparison to those of Korea and Brazil. India adopted an ISI strategy from the beginning. ISI was radically intensified with the Second Five-Year Plan 1956–61 with a focus on building a heavy industrial base reserved largely for the public sector. The growth rate of Net Domestic Product (NDP) from 1956–57 to 1965–66—i.e., over the Second and Third Five-Year Plans—was only 2.9%. Industry and manufacturing grew much more rapidly at 6.4% and 6.0%, respectively. Agriculture was the drag on overall growth, registering a rate of only 0.5% during this period.[19]

This phase of fairly rapid industrialization reached a crisis in the consecutive famine years of 1965–66 and 1966–67. The food imports necessitated by famine and the war with Pakistan in 1965 precipitated a foreign exchange crisis. The rupee had to be devalued in 1966. The mid- to late 1960s were recognizable as a crisis of the ISI model. However, unlike in Korea in 1962 and in Brazil since the late 1960s, the crisis was eventually resolved by intensified ISI to save foreign exchange.

Growth rates from 1966–67 to 1981–82 were 4.0%, 4.5%, 4.3%, and 3.1% for NDP, industry, manufacturing, and agriculture, respectively, revealing a slight slowdown in industrial and especially manufacturing growth rates compared to the earlier period. Agricultural growth picked up, however, due to the spread of the Green Revolution in the northwestern states. According to the World Bank, 1965–80 growth rates were 3.7%, 4.0%, 4.3%, 4.6%, and 2.8% for GDP, industry, manufacturing, services, and agriculture, respectively. Over 1980–90, GDP growth picked up to 5.3%, as did manufacturing growth to 7.1%. The percentage composition of GDP in agriculture/industry(manufacturing)/ services changed over 1965/80/92 from 47/22(15)/31 to 32/29(19)/39 to 32/27(17)/40, remaining static despite faster growth over 1980–90. The percentage distribution of the labor force in agriculture/industry/services changed only very slightly from 73/12/15 to 70/13/17 over 1965–80. Urbanization increased from 19% to 25% to 26% over 1965–80–92. Thus there was only a very small shift in the occupational distribution of the labor force and only a small rural–urban shift over 1965–92. The slowdown in industrial and manufacturing growth rates since the mid-1960s happened despite the increase over 1965–80 in the gross domestic savings rate from 15% to 21% and in the gross domestic investment rate from 17% to 21%. This was reflected in the rise in (worsening of) the incremental capital–output ratio.

The causes of slow industrial growth since the mid-1960s have given rise to a debate in India.[20] Ahluwalia has identified four causes of the slowdown: (1) slow growth of agricultural incomes; (2) slowdown in the rate of growth of public investment, especially in infrastructure, since the mid-1960s; (3) poor infrastructural management; and (4) the industrial policy regime, which created a high-cost economy. The last three causes are ultimately political. We do not go into the debate here, but we outline the industrial policy regime in brief.

The Indian solution to the crisis of the first phase of ISI was to go in for intensified blanket ISI. India's export performance was very poor compared to Korea, Brazil, and other NICs. Exports grew extremely slowly over 1965–73 at 8.1% per annum, compared to 50.3% for Korea, 33.3% for Brazil, 31.1% for Argentina, and double-digit percentages for all other NICs.[21] Over 1960–77, India's share in world exports fell from 1.04% to 0.48%; excluding minerals and fuels, it fell from 1.15% to 0.69%. India's share of total LDC exports over the same period fell from 4.81% to 1.92%. If we exclude the eighteen major oil exporters, it fell from 7.04% to 4.93%.[22] The average annual growth rate of exports over 1965–80 was 3.0%, and 6.5% over 1980–90. India's exports in 1992 were only $20 billion ($14 billion manufactures) compared to Korea's $76.4 billion ($71 billion manufactures) and Brazil's $36 billion ($20 billion manufactures). Despite this poor performance, the composition of exports shifted significantly in favor of manufactures, including those other than textiles. The composition of merchandise exports in terms of primary–machinery–other (textiles) shifted from 51/1/47(36) to 27/7/66(23) over 1965–90. The inability to export was a major factor in the relatively slow growth of Indian industry and the economy as a whole, as compared to Korea and Brazil. The former's growth was export-led, while the latter's brand of intensified ISI in the 1970s was not biased against exports as much as India's and, in fact, was accompanied by rapidly growing exports, especially of manufactures.

Capital goods production in India grew at a rapid pace during the Third Five-Year Plan—that is, at 19.7% per annum over 1961–65. It declined to 5.4% per annum in the 1970s, so that by 1978 its gross output was half Brazil's and barely larger than Korea's.[23] Indian capital goods ISI was aimed at self-sufficiency, the domestic supply ratio rising to a very high 84.3% by 1978. However, Indian capital goods were relatively uncompetitive, their 1978 export of $0.5 billion being barely a fifth of that of Brazil and Korea in 1979 and mainly to developing countries.[24] However, Indian capital goods production involved more complex products and until the later 1980s was more broad-based than Korea's and possibly Brazil's despite being less efficient, a feature cited by both critics and defenders of Indian ISI to support their respective arguments.

Despite this poor export performance and despite persistent trade deficits, India managed to avoid coming to the brink of default on its debts until the mid-1991 crisis. This is because India has had access to concessional finance from the World Bank group and until the 1980s avoided borrowing on commercial terms like Brazil, Korea, and other middle-income developing countries. The period since 1991 has been one of stabilization, structural adjustment, and slow growth.

As far as equity and basic needs aspects of such a strategy are concerned, the experiences of Korea and Brazil are a study in contrast. Income inequality was much worse in Brazil. There were also sharp contrasts in other social indicators. These contrasts can be explained to a great extent by the fact that South Korea had a thoroughgoing land reform in the early postwar period which reduced rural inequalities and laid the basis for rapid agricultural growth. Also, the government during the export-led growth phase after 1962 did not neglect agriculture, nutrition, health, and education. Brazil since 1964 has had neither such land reforms nor such strong basic needs policies.[25]

KEY DIFFERENCES IN DEVELOPMENT STRATEGIES AND THE ROLE OF THE STATE

There were key differences in the overall industrial strategy of the three countries in their responses to the crisis of the first round of ISI, their policies toward FDI and its importance for export-promotion, their policies toward foreign borrowing, and, lastly, the spheres and instruments of intervention of the state. These differences reflect the nature of the social–coalitional base of the state and have very important implications for the possibility of successful strategy-led development. This overview of broad strategy is, therefore, the necessary background for our detailed comparative study of the electronics industry as a case of the possibility of strategy-led development for competitiveness.

Differing Responses to the Crisis of the First Round of ISI

The alleged exhaustion of ISI is a notion that can be accepted for Brazil in 1964 only if we refer to a specific phase of ISI: that is, substitution of nondurable consumer goods like textiles. ISI continued throughout the period of rapid growth of 1965–80 in sectors like consumer durables, capital goods, and high-technology industries. In Korea, while ISI was no longer the principal stimulus, it did continue in a selective way. In its case, however, the narrowness of the domestic market and the balance of payments strain due to imports of most inputs made ISI more limited as a strategy and impelled an earlier shift to export-led growth. Nevertheless it should be noted that in the crucial capital goods and advanced technology sectors, the carriers of technological capability and competitiveness, selective ISI has been continued by both countries. As for India, Ahluwalia has noted that ISI was not exhausted and continued after the crisis of the mid-1960s with greater emphasis on backward vertical integration and self-sufficiency. The slowdown in growth rates was not due to the exhaustion of ISI.[26]

The Role of Foreign Direct Investment

Brazilian and Korean experiences have been markedly different as regards FDI by MNCs and its effect on exports and technology transfer, calling into question whether such investment is necessary for exports or capital goods production ("deepening" in O'Donnell's terminology).[27] MNC investment has played a major role in Brazil, and a relatively minor one in Korea.

FDI in Brazil, especially in manufacturing, rose during the late 1960s and 1970s, alongside large-scale state investment in industry and infrastructure which was strongly complementary to, rather than competitive with, MNC investment.[28] In contrast to Brazil, Korea followed a policy of tight restrictions on direct foreign investment while similarly undertaking heavy state investment in industry. Like Brazil, however, Korea also went in for heavy borrowing on international capital markets. For Korea, direct investment was only a small percentage of net capital inflow; 3.7% of net inflow over 1967–71 and 7.9% over 1972–76, while the figures for Brazil were 33.8% and 22.9% in the same periods. Korea's Foreign Capital Inducement Plan (1965) and its revisions gave numerous incentives to foreign investors. However, the access of foreign firms to the domestic market was restricted severely, and local content requirements in certain sectors forced domestic linkages. Prior screening procedures for FDI have explicitly sought to limit foreign investment in sectors where Korean firms can be competitive internationally. Joint ventures have been preferred except in Free-Trade Zones (FTZs). Of FDI in manufacturing, 61.8% had at least 50% Korean participation, and only 25.5% was 100% foreign.[29]

FTZs for assembly of labor-intensive products (such as the Masan Free Export Zone and two others) were not typical of FDI in Korea, accounting for only 10% of the total influx of foreign capital by 1975, mainly Japanese. As late as 1984, of the over $40 billion of long-term investment financed by foreign capital, less than $3 billion came from FDI, and that, too, mainly as joint ventures.

In the Indian case the exclusion of FDI was even more radical. The passage of the Foreign Exchange Regulation Act (FERA) in 1973 ordering foreign firms to dilute overseas equity to 40%, unless 100% export-oriented or in high-technology manufacturing, led to a sharp fall in the inflow of new FDI and, in fact, to a pullout by many foreign firms. It did not induce foreign firms to either export from India or bring in high-technology to retain controlling ownership.

MNCs and Exports

As for the relationship between FDI and exports, MNC subsidiaries have accounted for a large proportion of Brazilian exports, especially manufactured goods, while in Korea their role, like their weight in Korean industry, has been much smaller. In 1974, MNCs were responsible for about 43.3% of Brazil's manufactured exports, compared to 33.8% in 1967.[30] MNCs' participation in exports was pronounced in capital goods industries like automobiles, machine tools, and electrical machinery—i.e., in the more complex manufactures. Foreign firms were among the four largest exporters in eight product lines that accounted for 55% of the value of capital goods exports in 1980.

The MNC share in Korean manufactured exports was only around 15% in 1971 and 18.7% in 1978.[31] This was mostly in electronics assembly in the Free Export Zones, and later (after 1973) in chemicals and capital goods, thus appearing to resemble Brazil in that complex manufactures tended to be disproportionately MNC-dominated.

However, there were major differences of degree. Even in chemicals, electronics, and capital goods exports, the MNC share was much smaller than in Brazil. By 1979, the share of domestic firms in electronics exports, earlier the atypical case of MNC domination in Korea, was 45%, compared to 26% in 1970.[32] Capital goods exports, which are highly diversified, were overwhelmingly made by the *chaebol* (the Korean business conglomerates) and by state-owned firms.[33] In the case of India, MNCs accounted for only a very small proportion of Indian exports, manufactured and otherwise.

These figures reflect the growth of the technological capacity behind competitiveness, underlining the fact that *FDI was not absolutely necessary.* Licensing and outright purchase by indigenous private and state enterprise can be an effective way to acquire technological capacity. This was particularly so in the case of India, where there was a marked policy hostility toward foreign equity participation and toward technology import in general. The "deepening" of the industrial structure into heavy, chemical, and capital goods industries and electronic components (as against assembly of imported ones) after 1973 was carried out by state enterprise (steel, heavy machinery, shipbuilding) and by the *chaebol* (electronics, electrical machinery, engineering, transport equipment) with state backing. Technology acquisition was mainly through licensing followed up by local effort.

What can we conclude from all this about FDI and exports, backward integration into capital goods, and technological capacity? Despite Korea's much smaller FDI share in both manufacturing production and exports, including in the crucial capital goods sector (again in both production and exports), Korean growth rates of GDP, manufacturing production and manufactured exports, including capital goods exports were higher than those of Brazil and India from the mid-1960s. By 1990, Korea's manufactured exports were in absolute value four times those of Brazil. We can conclude, therefore, that FDI predominance was not an essential condition for export-led growth, capital goods production, and technological capability acquisition.

Foreign Borrowing, the Balance of Payments, and the Debt Crisis

Economic liberalization policies such as lowering of trade barriers and devaluation to realistic exchange rates often lead to a surge of imports before the export growth materializes, resulting in a temporary worsening of the balance of payments (the J-curve effect in the case of devaluation). It has been shown that trade and financial liberalization can even be contractionary.[34] This aside, does a drive to develop capital goods and other technologically complex industries set up through heavy state investment and/or state-controlled preferential credit, backed by foreign loans rather than FDI, worsen the balance of payments and lead to a debt crisis, or can it create a comparative advantage and produce sustained export-led growth? Alternatively, does the creation of a heavy industrial base through borrowing rather than FDI simply mean another kind of harness? The Indian ISI model in the 1970s also depended on foreign concessional assistance rather than FDI. What is crucial in all strategies is the competitiveness of the new industry created.

Korea avoided Brazil's degree of dependence on MNCs but went in for heavy foreign borrowing like the latter, a large part of this being by state enterprises in the new heavy and chemical industries in the 1970s. Brazil's foreign debt ballooned from $17 billion in 1974 to almost $54 billion in 1980 and about $110 billion in 1986, almost all the movement due to borrowing from private international sources of finance. In the 1980s the need to service incurred debts alone required further borrowing, and debt–service ratios were over 60% in 1980–82. A devastating recession attended by a fall in GDP and in exports, with recovery only in 1985, followed the 1979–80 oil shock which knocked a hole in Brazil's balance of payments. The debt crisis was finally overcome only by the early 1990s, after a decade of stagnation, leading to the restoration of international lending, following the long-overdue stabilization and structural adjustment policies of the Collor administration.

Korea, however, while also going in for foreign private borrowing in the 1970s, had a rather different experience. Korea's debt–service ratio moved in the opposite direction to those of other debtors, going from 20% in 1970–72 to 13% in 1980–82 due mainly to its success in raising exports, despite a very high debt/GNP ratio of 44% by 1983. The shock of 1979–80 was overcome much more quickly by Korea. The economy stagnated in 1979–80, with capital goods production taking a steep drop of 15% after skyrocketing growth.[35] The government adjusted by increasing foreign borrowing, devaluing the currency, reducing public investment and real wages (all orthodox IMF-type measures also applied in Brazil), and concentrating credit on export-production sectors. GDP growth resumed in 1981, inflation declined dramatically, and real export growth accelerated to 10%. The current account deficit was cut by two-thirds, and the growth of external debt slowed down. Access to foreign financial markets in the late 1980s was normal. The long-term debt/GDP ratio by 1990 was down to 9.5%, compared to 29% in 1980, and the long-term debt service/export ratio was down to 9.6% compared to 14% in 1980. *The key to the contrast with Brazil is in Korea's greater success in raising exports.* Behind this success the most important factor was technological capability development over the years; policies were followed specifically for the acquisition of such capacities and the promotion of internationally competitive firms.

India virtually excluded FDI but depended on foreign concessional financing especially from the World Bank for bridging its persistent trade deficits and its domestic resource shortfalls, though decreasingly so over the Fourth (1969–74) and subsequent Five-Year Plans. This changed in the second half of the 1980s, with increased resort to borrowing on commercial terms leading to a ballooning of the long-term debt/GDP ratio from 11% in 1980 to 27% by 1991 and of the long-term debt service/export ratio from 8.5% in 1980 to 24% by 1991. However, it avoided coming to the brink of default until the crisis of mid-1991, leading to IMF-supported stabilization and still-ongoing structural adjustment. The new policies underline the past mistake of neglecting competitiveness.

We can conclude that debt-based growth is bound to peter out in a balance of payments crisis at some point unless a strong, indigenous, internationally competitive industro–technological base is created during the years of borrowing. This did not happen to the required extent in Brazil and India if one uses manufactured exports as a measure, though there certainly was considerable technological progress.

The Role of the State: Spheres and Instruments of Intervention

Were the Korean and Brazilian states doing what had been said about Korea: "Korea provides an almost classic example of an economy following its comparative advantage and reaping the gains predicted by conventional economic theory."[36] It is now accepted that even in Korea, let alone Brazil, export-led growth entailed an economic liberalization that was highly selective and was accompanied by extensive and decisive state intervention in the economy in at least three ways: (1) disguised or open protection for the new round of ISI in capital goods as well as for the domestic consumer and intermediate goods markets; (2) massive state investment in capital goods and other heavy, chemical and advanced industries, and (3) control over both private and state enterprise investment through control over the banking and financial institutions.

Indeed, in both Brazil from the late 1960s on and Korea from 1973 on, public industrial investment was massively stepped up and selective ISI took place in capital goods, chemicals, and other technologically complex industries. These were not the only "aberrations" in economic liberalization in Brazil and Korea. Lowered tariff barriers on imports have led to less openness than has generally been believed. What has attracted the attention of liberal economists in the case of Korea are steps like high interest rates to curb the growth of money supply and raise savings in terms of financial assets, devaluation, and the elimination of many domestic economic controls, which look like IMF-type orthodoxy. However, Korea had a list system on imports for over two decades since the turn to export-led growth, varying from three to five lists at different times and ranging from automatic imports to banned items. The list and tariff systems have been biased in favor of raw materials and intermediate inputs for export-oriented industries while maintaining a significant degree of protection for consumer goods and other finished goods sold in the domestic market. There were also significant nontariff barriers. Looking at tariffs alone gives us a more "liberal" trading picture than was actually the case over most of the past three decades.

In Brazil, the more significant import restrictions were nontariff barriers, especially from 1972 on. Direct controls over imports, especially capital goods, were also exercised by state enterprises. Hence we can see that lowered tariff barriers and import-intensive industrialization in Brazil and Korea went hand-in-hand with a considerable degree of direct and indirect protection of the domestic market, and not only for capital goods. As for guidance of industrialization via control over the capital market, the orthodox economists' focus on the relationship between trade policies and growth has tended to ignore the role of capital market controls in state guidance of industrialization.

The Korean government promoted export-oriented industries through indirect intervention by way of highly differentiated interest rates, export credits, export-oriented import-financing credits, and foreign exchange allocation preference. The regime has controlled the banking system since 1961. From 1964 on, financial institutions concentrated portfolios on government-designated, export-oriented industries and product categories.

State control over credit came to play three mutually reinforcing functions: (1) as a means to intervene in the internal decisions of private and state-owned industrial enterprises, crucially in investment decisions; (2) to subsidize the initial entry of firms into foreign markets; and (3) to cement a business–state alliance in which the latter was the stronger partner and exercised control.[37] The industrial financing policy was reminiscent of that of Japan, with major decisions on choice of industrial sector, subsector, and product types taken on the basis of long-term objectives, and in disregard of short-term efficiency as indicated by existing prices. Adequate finance and the ability to channel it selectively to potential export industries, and in them to promising firms, was critical to the success of export-led growth.

In Brazil, too, the government exercised considerable control over finance. Public sector institutions dominated the financial system. However, state financial control over the industrialization process during the economic liberalization phase after 1964 differed in crucial ways from that in Korea, with weighty consequences for economic strategy. Total state-controlled savings in Brazil were about 64% of total savings by 1976. State banks controlled 50% of savings in commercial banks, and an even larger share of loans. Five out of the top twenty commercial banks were government-owned. By 1974, Banco do Brasil (60% state-owned) held 37.1% of all funds on deposit in Brazil's fifty biggest banks, and all state-owned banks held 55% of total deposits with the banking system. State-controlled banks financed both state and private enterprise. The National Housing Bank (BNH) and the National Economic Development Bank (BNDE) financed a large part of private investment. BNDE and a constellation of development banks owned by Brazilian states provided 100% of project financing in cruzeiro resources. The banks' Special Agency for Industrial Finance (FINAME) program provided low-cost loans to finance domestically produced equipment purchases. Such public sector financing has been at the heart of the growth of capital goods production in Brazil since 1974.

The role of the Brazilian state and the functioning of its control in the 1960s through the 1980s were quite different from those of Korea. Selective ISI during the "miracle" years of 1968–74 did not protect industries with a view to the possibility of building up an international competitive capability. State control over the major part of the banking system was not used to channel credit selectively to promote the manufacture of products in which Brazil could become globally competitive using the right technologies, having scale economies required for export-capable efficiency levels, and following up on technology development. This was despite the fact that the state not only had control over both the banking system and hence external financial flows to state enterprises, but considerable say in the internal decisions of these firms.

The kind of business–government joint planning that took shape in Korea did not materialize in Brazil. On the contrary, short-term market criteria determined investment decisions. The boom of 1968–74 was based on consumer durables for the limited upper-income domestic market without attention to internationally optimal economies of scale and choice of product and technology appropriate to their factor endowments, and hence, comparative advantage. It was also based on government-financed construction, often of a luxury character. The take-off into capital goods after 1974 was again without attention to scale and technology. Choice of product, scale, and technology was determined by immediate domestic demand criteria and, to a lesser extent, by balance of payments considerations after 1973, which dictated backward integration to reduce import dependence. What is surprising is that despite the creation of a large public enterprise sector especially in capital goods and heavy industries, these enterprises developed in a haphazard way according to market forces, and they were *not controlled accord-*

ing to any strategic plan to develop certain industries as internationally competitive.

As far as India was concerned, the all-pervasive role of the state in the economy continued after the mid-1960s, including the complex and cumbersome industrial and import licensing systems. What was different was the decline in the rate of growth of public investment. Ahluwalia has identified this as an important factor in the slowdown in the rate of industrial growth after the mid-1960s, especially in the capital goods sector. Another important difference from the earlier period was the nationalization of the major commercial banks in 1969. The government now had the potential to influence the direction of both state and private industrial investment even more than before. However, Indian industrial policy took a peculiar turn, accentuating its already unique form of control over the private sector—the industrial licensing system. We shall deal with the Indian overall industrial policy regime in detail in Chapter 5. Here it is sufficient to say that, unlike in Korea, state control of banking in India was not used to promote export-oriented industries. Indeed, it was not used to promote even efficient ISI. For ultimately political reasons, it came to be used to promote blanket ISI regardless of comparative advantage or even strategic significance.

Crucial Differences in State Intervention Patterns

If Korea, Brazil, and India all had interventionist states and intervened in broadly the same three spheres discussed above, then why did their strategies end up with substantially different results? Korea was able to do much better than Brazil and India on the export front in terms of both growth rates and absolute value of exports (even given its smaller GNP) and did not sink like Brazil into a balance of payments and debt crisis following the second oil shock of 1979–80. This was due to the greater competitiveness of its firms. Although it can possibly be argued that the difference lay crucially in the policy framework, like in trade policy, this does not really answer the question since a protectionist trade policy did not hinder the development of exports in Korea. Offsetting policies like export-conditional credit or subsidization which preserve export incentives are possible as in Korea. The questions that then arise are: What were the specific policy differences in state guidance that led to greater industrial competitiveness in the Korean case? To what uses were the instruments of import barriers, public investment, and financial controls over state and private enterprises put?

The crucial difference in the nature of state intervention that separates the Korean from the Brazilian and Indian cases through the three decades of its export orientation is the single-minded concentration on fostering internationally competitive industries. For Korea it was not enough to integrate backwards into more advanced industries by creating a capital goods base. The latter had to be internationally competitive within a short time, if not from the beginning, for such creation to be judged desirable or at all worthwhile. The goal was not just self-sufficiency and import-independence in capital goods, chemicals, electronics, and other advanced industries. This single-minded determination shows up in the kinds of state intervention Korea undertook. It does not show up so clearly in the case of the Brazilian and Indian states. The details of the differences in the state's industrial strategy and the causes of these differences are highlighted in our detailed study of the electronics case in the following chapters.

We can reasonably conclude from the discussion above that for both import-intensive, export-led growth as well as even "efficient" ISI to be sustained over the long-term certain stringent necessary conditions have to be met. Exports have to grow fast enough to be able to overcome the negative effect on the balance of payments of opening up the economy, or of persistent trade deficits due to initial uncompetitiveness, and the consequent jump in external debt. For the export momentum to be maintained it is necessary to acquire and indigenize the technologies required for a fully competitive industrial base, and to create an indigenous technology-generating capability in industry. These essential conditions are not necessarily exhaustive. Developing such a competitive capability requires an appropriate industrial policy framework and, in turn, a conducive political economy, which obtained in Korea and not in Brazil or India.

ELECTRONICS AND THE POLITICAL ECONOMY OF STRATEGY-LED DEVELOPMENT

A comparative study of the political economy of development of the electronics industry in India, Brazil, and Korea is critical for an assessment of the possibilities of successful strategy-led development, resulting in competitiveness, for three reasons:

1. We explained earlier why India, Brazil, and Korea are key contrasting cases for a comparative study of industrial strategy in the Third World in general. With regard to electronics, all three countries have since the end of the 1960s made efforts to promote a local electronics manufacturing base and with widely differing results. By all conventional indicators, Korea has been the most successful, with Brazil a distant second, followed by India. This is the case despite India starting more or less on a par and having a head start in spotting opportunities and planning for the industry and, in fact, in initiating local manufacture in many products well before Korea and in many cases before Brazil. All three countries entered a new phase of electronics policy and development in the early 1990s.
2. The electronics industry is a critical case because it is a technologically dynamic industry whose promotion inescapably involves all the policy dilemmas of industrial strategy: ISI versus export orientation, foreign versus domestic firms, public versus private sectors, technology import versus domestic development, choice of product for specialization, and timing of entry and phasing.
3. An electronics industry promotion strategy involves dilemmas not just of industrial policy but of politics. It does so precisely because of the policy choices

outlined above, which pit different interests against each other and/or the state over decisions about such controversial issues as FDI versus domestic capital, and public versus private sectors. But it also does so in a unique way that throws into relief the interests of the state elite as distinct from the components of its social–coalitional base. Electronics is the quintessential high-technology industry. It is widely felt to be a "strategic" industry in the sense of being vital to competitiveness in other industries (including defense) in an age of computerization and networked communications, and of the application of microelectronics and its technologies to economic activity of all kinds.

Why did India fall behind despite being roughly on par to begin with or even enjoying a head start over Korea and Brazil? A comparative study throws light on the political and economic requirements for possessing strategic capacity and for effective strategy-led development in the Third World. It leads to much greater specificity about the conditions for, and possibility of, replicating the development of Korea and other successful NICs.

In what follows we analyze the political economy of Indian electronics strategy from around 1970 to 1992, including the new directions since then, comparing it with Korea and Brazil, in an attempt to extend, refine, and develop further the concept of strategic capacity by taking the perspective "from behind" of a country that lost its initial edge over time to lag behind by most indicators.

NOTES

1. Bela Balassa et al., *The Structure of Protection in Developing Countries* (Washington, DC: International Bank for Reconstruction and Development, 1971); Anne O. Krueger, *Foreign Trade Regimes and Economic Development: Liberalization Attempts and Consequences* (Cambridge, MA: Ballinger, 1978); I. M. D. Little, Tibor Scitovsky, and Maurice Scott, *Industry and Trade in Some Developing Countries: A Comparative Study* (London: Oxford University Press, 1970); Larry E. Westphal, "The Republic of Korea's Experience with Export-led Industrial Development," *World Development,* 6, no. 3 (March 1978).

2. See Frederic C. Deyo (ed.), *The Political Economy of the New Asian Industrialism* (Ithaca, NY: Cornell University Press, 1987), especially the essays by Johnson, by Haggard and Cheng, and by Deyo; Stephan Haggard, and Chung-In Moon, "The South Korean State in the International Economy: Liberal, Dependent or Mercantile," in John Gerard Ruggie (ed.), *The Antinomies of Underdevelopment: National Welfare and the International Division of Labor* (New York: Columbia University Press, 1983).

3. For the characterization of Brazil's dominant class coalition as such a triple alliance, see Haggard and Moon, in Ruggie, *Antinomies,* for the use of the term "mercantilist" to describe Korea; Peter B. Evans, *Dependent Development: The Alliance of Multinational, State and Local Capital in Brazil* (Princeton, NJ: Princeton University Press, 1979).

4. For examples from these studies, see Balassa, *Structure of Protection*, Krueger, *Foreign Trade Regimes*, Little, Scitovsky, and Scott, *Industry and Trade*,

Jagdish N. Bhagwati, *Foreign Trade Regimes and Economic Development: Anatomy and Consequences of Exchange Control Regimes* (New York: Columbia University Press, 1976).

5. On tariffs, FDI, and immiserizing growth, see Jagdish N. Bhagwati and Richard A. Brecher, "National Welfare in an Open Economy in the Presence of Foreign-owned Factors of Production," *Journal of International Economics,* 10 (1980); Richard A. Brecher and Carlos F. Diaz-Alejandro, "Tariffs, Foreign Capital and Immiserising Growth," *Journal of International Economics,* 7 (1977): 317–322.

6. Peter B. Evans, "Class, State and Dependence in East Asia: Lessons for Latin Americanists," in Deyo, *Political Economy*, p. 205.

7. Frederic C. Deyo, "Coalitions, Institutions, and Linkage Sequencing—Toward a Strategic Capacity Model of East Asian Development," in Deyo, *Political Economy*.

8. Chalmers Johnson, "Political Institutions and Economic Performance: The Government–Business Relationship in Japan, South Korea and Taiwan," in Deyo, *Political Economy*, p. 142.

9. Johnson, in Deyo, *Political Economy*, p. 145.

10. Robert Wade, *Governing the Market: Economic Theory and the Role of Government in East Asian Industrialization* (Princeton, NJ: Princeton University Press, 1990), p. 28.

11. Alice Amsden, *Asia's Next Giant: South Korea and Late Industrialization* (New York: Oxford University Press, 1989); Ashoka Mody, "Institutions and Dynamic Comparative Advantage: The Electronics Industry in South Korea and Taiwan," *Cambridge Journal of Economics*, 14 (1990): 291–314.

12. Peter B. Evans and Paulo Bastos Tigre, "Going Beyond Clones in Brazil and Korea: A Comparative Analysis of NIC Strategies in the Computer Industry," *World Development,* 17, no. 11 (1989): 1761.

13. Peter B. Evans, "Indian Informatics in the 1980s: The Changing Character of State Involvement," *World Development,* 20, no. 1 (1992): 2.

14. Evans, "Indian Informatics," p. 13.

15. Baldev Raj Nayar, "The Politics of Economic Restructuring in India: The Paradox of State Strength and Policy Weakness," *Journal of Commonwealth and Comparative* Politics, 30, no. 2 (1992): 147.

16. Nayar, "Politics of Economic Restructuring," p. 148.

17. All figures quoted in this section are from, or are computed from, World Bank, *World Development Report*, Tables, 1990–95, and World Bank, *Trends in Developing Economies 1992*, unless otherwise specified, and have been cross-checked for consistency across annual issues.

18. For capital goods exports, see Daniel Chudnovsky, Masafumi Nagao, and Staffan Jacobsson, *Capital Goods Production in the Third World: An Economic Study of Technology Acquisition* (New York: St. Martin's Press, 1983), Table 3.1, p. 96.

19. See Isher J. Ahluwalia, *Industrial Growth in India: Stagnation since the Mid-Sixties* (New Delhi: Oxford University Press, 1985), p. 3, Table 1.1.

20. For a detailed study of industrial stagnation since the mid-1960s, see Ahluwalia, *Industrial Growth*; for a summary of the debate, see Ashutosh Varshney, "The Political Economy of Slow Industrial Growth in India," *Economic and Political Weekly*, 19, no. 35 (1984).

21. Ahluwalia, *Industrial Growth*, Table 6.2, p. 118.

22. Ahluwalia, *Industrial Growth,* Table 6.1, p. 117.

23. Chudnovsky, Nagao, and Jacobsson, *Capital Goods*, p. 98; p. 96, Table 3.1.

24. Ibid., p. 96, Table 3.1.

25. For analyses of Korean and other East Asian countries' growth-with-equity patterns, see World Bank, *The East Asian Miracle: Public Policy and Economic Growth* (New York: Oxford University Press, 1993).

26. Ahluwalia, *Industrial Growth*, pp. 166–167.

27. Guillermo O'Donnell, "Reflections on the Patterns of Change in the Bureaucratic–Authoritarian State," *Latin American Research Review*, 8, no. 1 (1978): 3–38.

28. Ronaldo Munck, "State Intervention in Brazil: Issues and Debates," *Latin American Perspectives*, 4, no. 4 (1979): 22. Munck takes a traditional dependency view of Brazilian development, emphasizing the state-MNC alliance rather than seeing the state as nationalistic.

29. See Haggard and Moon in Ruggie, *Antinomies*, and Hubert Schmitz, "Industrialization Strategies in Less Developed Countries: Some Lessons of Historical Experience," *Journal of Development Studies*, 21, no. 1 (1984), and for the above figures.

30. Deepak Nayyar, "TNC's and Manufactured Exports from Poor Countries," *Economic Journal*, 88 (March 1978), p. 62, Table I, and p. 80.

31. Ibid., p. 62.

32. Korea Development Bank, *Industry in Korea* (1980), p. 99.

33. Larry E. Westphal, Yung W. Rhee, Linsu Kim, and Alice Amsden, "Republic of Korea," *World Development*, 12, no. 5/6 (1984): 515–518; Sanjaya Lall, "India," *World Development*, 12, no. 5/6 (1984): 543–544.

34. Paul Krugman and Lance Taylor, "Contractionary Effects of Devaluation," *Journal of International Economics*, 8 (1978); Edward F. Buffie, "Financial Repression, the New Structuralists, and Stabilization Policy in Semi-Industrialized Economies," *Journal of Development Economics*, 14 (1984).

35. Chudnovsky, Nagao, and Jacobsson, *Capital Goods*, p. 95.

36. Westphal, "The Republic of Korea's Experience," p. 375.

37. Haggard and Moon in Ruggie, *Antinomies*; Schmitz, "Industrialization Strategies"; Leroy Jones and Il Sakong, *Government, Business and Entrepreneurship in Economic Development: The Korean Case* (Cambridge, MA: Harvard University Press, 1980).

2

The Evolution of the World Electronics Industry: Innovation, Technological Regimes, Associated Market Structures, and Policy Implications for Developing Countries

RELATING THE TECHNICAL TO THE ECONOMIC AND POLITICAL

A window of opportunity opened for developing countries at the end of the 1960s. Electronics, a technologically complex industry, until then the preserve of the most advanced developed countries, began to become accessible to developing countries. Changes in the technology and market structure of the industry lowered the technological and financial barriers to entry into the industry for new firms and countries. This tendency affected more and more segments of the industry over the following decade, making it possible for countries like India, Brazil, and Korea to launch ambitious programs for local electronics manufacture.

To understand why the barriers to entry fell, thereby opening windows of opportunity, it is necessary to understand the technological and world market-structural changes in the industry over the past three decades. Essentially, what happened was that revolutionary innovations in the most basic electronic components—semiconductors—radically lowered the cost of electronic equipment even while boosting its power and versatility. These innovations also simplified production processes, which also improved the power and performance of components and equipment, leading to the emergence of labor-intensive production stages that were relocated in developing countries. Taken together, these changes lowered financial and technological entry barriers for smaller and less sophisticated firms in both developed and developing countries. In this chapter we explain the changes that opened up new opportunities in electronics for developing countries.

We have organized the material as follows, avoiding unnecessary technical and historical detail. (1) In the first section we explain what the electronics industry consists of, focusing on the critical electronic components—semiconductors.

(2) In the next section, we give an account of the landmark innovations in the semiconductor industry over the past four decades that have led to the lowering of entry barriers, spawned a host of new firms, and thereby reshaped market structures worldwide. The focus here is on the supply, demand, and public policy factors in the United States, Japan, and Western Europe in the evolution of the semiconductor industry. (3) The following section examines the implications for the electronic equipment industry (computers, communications, and so forth) of these innovations and market-structural changes at the component (semiconductor) level. It highlights the lowering of financial and technological entry barriers and the simplification of production processes in these segments of electronics and the consequent changes in market structures due to the host of new entrants, including those from developing countries. It also elaborates on the technological convergences across the electronics complex of industries. (4) The final section examines the policy and political implications of various strategic options for developing countries.

THE ELECTRONICS COMPLEX OF INDUSTRIES: A BRIEF TECHNICAL INTRODUCTION

The Subdivisions of the Electronics Complex

The importance of electronics, especially applications of electronics to the rest of the economy, can hardly be overemphasized today, even in developing countries. Traditional industries like steel, machine tools, and textiles cannot be efficient and competitive without electronic process-control systems and instrumentation to monitor inputs, outputs, and process parameters and thereby control costs and quality. Nor can industrial management do without data-processing equipment. Nor can the economy as a whole do without an efficient tele- and datacommunications infrastructure. And modern weapons systems and military command and control systems depend vitally on microelectronics-based computer and telecommunication (henceforth telecom) systems. Even consumer electronic products like radio and television play a vital role in mass communication for development, and also as the principal source of demand for components at the early stages of development of the component industry as in the experience of Europe and Japan.

Electronics is a heterogeneous complex of industries based on components of a common technology.[1] Broadly speaking, it can be subdivided into:

1. *Consumer electronics:* radio, television (TV) receivers, audio and video-cassette recorders (VCRs), calculators, and similar products.
2. *Computers and Office Automation or Electronic Data-Processing (EDP) Equipment:* computer hardware and software, including the whole range of peripherals like printers, modems, monitors, and consumables; software is an increasingly critical segment.

3. *Telecommunication Equipment:* switching, transmission and terminal (usually on subscriber premises) equipment; these range from large telephone exchanges to radio and TV transmission towers to simple telephone instruments; switching equipment refers to systems like telephone exchanges which switch incoming electrical signals to other channels; transmission includes all kinds of transmitters and receivers (other than radio and television) and cables; terminal equipment includes telephones, teleprinters, and data terminals.
4. *Control Instrumentation and Industrial Electronics (CIIE):* an extremely heterogeneous category that includes industrial process-control systems and instrumentation, automotive electronics, medical electronics, and analytical, measuring, and test equipment of all types.[2]
5. *Aerospace and Defence Electronics:* electronic guidance and control systems for aircraft, missiles, and satellites, including ground-based systems such as radar, as well as for all types of military equipment and systems on land and sea and in aerospace.
6. *Components:* the building blocks of all electronic equipment; technological innovation in components, especially in semiconductor devices, has been extremely rapid and has revolutionized the electronics industry at the equipment and systems levels every few years.

Computers and telecommunications have been tending to converge since the early 1980s. Process-control systems, much used in defense and aerospace electronics, are a good example of this phenomenon. Such systems consist of networked special-purpose computers with particular communication protocols governing data transfer between them. This applies equally to office automation, military command systems, or scientific data networks. Examples are microcomputers embedded in weapons systems, as, for example, in the nose-cones of missiles. Process-control equipment of the current generation—that is, distributed digital control (DDC)—consists of a number of mini- and microcomputers networked together and connected to nodal large computers, each small system monitoring the flows of inputs, outputs and other parameters of a particular step in, for example, the manufacturing process. Each particular machine is controlled by the computer monitoring it and feeding back to it. The manufacturing process can thus be controlled "distributedly." In the 1990s, industrial control and automation are tending to become more decentralized and "intelligent," with the spread of artificial-intelligence applications, machine vision, and the like. Client–server architectures in information systems for office automation are analogous.

All electronic equipment industries are based on electronic components and it is components which have been at the cutting edge of technological advance, making possible advances in equipment and systems. It is therefore absolutely necessary to understand developments in the components industry, especially in the semiconductor industry, which we review briefly below.

Electronic components can be divided into passive and active components. Passive components—like resistors, capacitors, connectors, and relays—resist the flow of electricity, act as connections in a circuit, store charge, and relay signals.[3] Active components switch, modulate, amplify, and rectify electrical signals. They consist of electron tubes (both gas-filled and vacuum) and semiconductor devices. Electron tubes can be of many types, the best known being the TV picture tube, which projects the TV image onto the screen. However, there are myriad types of receiving and transmission tubes in telecommunications, radar, radio, and TV reception and transmission.

The World of Microchips: A Brief Introduction to Semiconductors

Semiconductor devices are the most critical and technologically revolutionary electronic components. Semiconductors are based mainly on materials like silicon (and in the near future will be increasingly based on gallium arsenide), which are either electrical conductors or insulators: at low temperatures they are insulators but under certain temperature conditions—and when "doped" by the "diffusion" in them of certain impurities such as phosphorus or boron—they become conductors.

Semiconductors can be classified into three major product groups: discrete devices, optoelectronic devices, and integrated circuits (ICs)—silicon "chips/microchips". In the first category are diodes and transistors. Included in the second are light-emitting diodes and liquid crystal displays, as for example, in digital watches and other display devices. It is the third category, that of ICs or "chips" or "microchips" in common parlance, that are responsible for a continuing technological revolution. ICs are tiny wafers of silicon, no more than a thumbnail in size, that have etched into them all the components of the circuitry of an entire device. There may be hundreds of thousands of components (such as transistors) on a single chip. These are immensely more powerful and versatile than discrete devices such as diodes and transistors, which are but single elements of an electrical circuit.

ICs can be classified by device type into (1) linear ICs, (2) digital ICs, and (3) hybrid ICs. Linear ICs are technologically less complex and are used mainly in analog electronic equipment such as radios, televisions, and consumer electronics. Digital ICs are the most complex. They can be categorized as logic chips, microprocessors, and memory chips. All of these are used in computers and other microprocessor-based systems such as process-control and telecom digital switching equipment. Logic chips are the ICs that do the arithmetical operations in computers—the actual computations.

Microprocessors are digital ICs that can perform all the functions of the central processing unit (CPU) of a computer. These functions are the arithmetical, the memory, and the control functions and are performed by the arithmetical, memory, and control units, respectively. The control unit regulates the flow of data from memory to logic units and vice versa. Microprocessors are extremely versatile chips that combine the functions of all three units of the CPU, dispensing with the need for separate logic, memory, and other chips. The microprocessor was invented in 1971 and made possible smaller and more powerful computers. However, independent logic chips are still used for large computers and other specialized powerful systems. Microprocessors are classified by the length of "words" processed at a time, that is, 4-bit, 8-bit, 16-bit, 32-bit, and so on.

Memories are ICs that can store information, measured in bits (binary digits). Types include random-access memories (RAM), read-only memories (ROM), erasable programmable read-only memories (EPROM) and electrically erasable programmable read-only memories (EEPROMs), and flash memories. Dynamic RAMs (DRAMs), the main memory chips used in computers, are the most important.

IC devices can also be categorized by technology. Transistors and ICs can be classified either as bipolar or metal-oxide semiconductor (MOS) technologies. Bipolar devices (both transistors and ICs) are much faster than MOS devices. MOS devices, however, have other advantages. They have higher circuit density, can have higher levels of integration (more components packed onto a chip), and have lower power consumption and heat generation. Bipolar ICs are typically used in telecom switching equipment, radar, image-processing, and other systems where very high-speed signal processing is required. MOS devices, especially complementary MOS (CMOS), are used in computer memories and microprocessors where high levels of integration and miniaturization are required. BiCMOS devices are a hybrid of the two technologies that attempt to get the best of both.

Semiconductor devices can also be classified by level of customization into standard, semicustom, and custom devices (or chips). Standard chips (such as the computer memory chips, DRAMs) are mass-produced, low-priced, and have wide application. They are also referred to as commodity chips. There are major economies of scale in their manufacture; volume and yield are the key to price-competitiveness, as demonstrated by the success of Japanese firms. Custom chips are designed for specific purposes and users (e.g., military and space applications) and are made in small volumes and are higher-priced. Semicustom chips facilitate the rapid and economical design and manufacture of custom chips. All these (i.e., custom chips and semicustom chips) are also called application-specific integrated circuits (ASICs).

The Stages and Economics of Semiconductor Manufacturing

Semiconductor manufacturing operations can be divided into five or six main stages, each with several steps. These are (1) the wafer-manufacturing stage; (2) the design stage; (3) the wafer-fabrication stage; (4) the assembly stage; (5) the final testing stage; and (6) the packing and shipping stage. The first three stages typically involve several steps each and are highly capital-intensive and skilled-manpower-intensive. So is the final testing stage, which is highly automated. These stages have usually been located in developed countries. The labor-intensive assembly stage, on the other hand, has been relocated in developing countries since the late 1960s to save on labor costs. These relocated assembly operations have usually been 100% MNC-owned and 100% export-oriented, the assembled chips being shipped back to the headquarters plants for final testing, packing, and shipping. However, since the early 1980s the assembly stage has increasingly become automated.

Semiconductor manufacturing has become increasingly capital-intensive. This is mainly due to steeply rising design and wafer-fabrication costs. In the design stage, increasing integration in the Large Scale Integration (LSI) and Very Large Scale Integration (VLSI) eras (i.e., over 100,000 components in a chip) implies increasing use of computer-aided design (CAD), which is extremely software- and knowledge-intensive. In the wafer-fabrication stage, photolithography (the drawing of electronic circuit lines on the silicon wafer by a fine beam of light) has been the main factor in raising capital costs as increasing integration and miniaturization of ICs have led to finer-line geometries now reaching lower submicron widths.

Unit costs and hence price-competitiveness in the semiconductor industry have been vitally affected by economies of scale and economies of learning. Economies of scale are operative in the circuit design and wafer-fabrication stages and are of increasing importance in the assembly and final testing stages as these become automated. Economies of learning, or learning effects, also play a crucial role in competitiveness through their effect on yields and incremental innovation and, in turn, on circuit design, integration, and miniaturization. Yields—the percentage of good units emerging from each production run—are important at the wafer-fabrication stage, the lowest-yield stage. Two technological bases for declining costs and prices over the past three decades have been (1) the learning curve in production, and (2) increasing density. The first applies at the level of specific products; the second is a phenomenon across the product spectrum. The first, and indirectly the second, are related to production volumes and hence economies of scale.

LANDMARK INNOVATIONS, TECHNOLOGICAL REGIMES, AND THE DECONCENTRATION OF THE SEMICONDUCTOR INDUSTRY

Technological Regimes and Changes in the Market Structure of the Semiconductor Industry: Supply, Demand, and Public Policy Factors in the United States, Europe, and Japan

Semiconductor technological innovation can be categorized into three eras or technological regimes: the transistor, the IC, and the microprocessor (LSI and VLSI) regimes. The first lasted from 1947, when the transistor was invented at Bell Labs, until 1961, when the first ICs were commercially produced. The second lasted about a decade, from 1961 to the advent of the microprocessor, developed at Intel corporation in 1971. This advance inaugurated the third and still current technological regime.[4] Technological regimes propel technological change in certain directions by influencing the R&D processes of firms. In the transistor era, the technological trajectory was toward device performance and reliability; in the IC era, toward miniaturization; and in the LSI and VLSI era, toward increasing function density and miniaturization.

The market structure of the semiconductor industry in the United States, Western Europe, and Japan evolved in response to these three key technological innovations and the technological regimes they gave rise to, but in significantly different ways. In the process, Western Europe, which was technologically equal to the United States in the early 1950s, lost ground in the 1960s and 1970s. Japan, a technological follower, caught up in the late 1970s and 1980s, although the United States seems to have pulled ahead again in the mid-1990s.

Following Malerba, we can analyze the evolution of the industry by looking at (1) supply, (2) demand, and (3) public policy factors. Supply factors include the organization of production and R&D, the structure of the industry, and firm strategies. Organization of production and R&D divide firms either into vertically integrated producers of both semiconductors and electronic final goods, or merchant vendors of semiconductors only. Industry structure refers to both concentration by market share and composition by strategic groups of firms. These distinctions differentiate between vertically integrated and merchant producers, and between major innovators and minor or incremental innovators. Firm strategies refer to major innovation strategies and minor or incremental innovation strategies. Demand factors include the magnitude, composition, and growth of demand. By composition is meant the final goods (equipment segment) shares in semiconductor demand. Demand factors overlap with public policy factors insofar as the latter create government (military, space) demand for semiconductors. Public policy factors, some of which overlap with supply factors, include six major sets of policies which are, following Wilson, Ashton, and Egan: government procurement, R&D funding, and trade, tax, antitrust, and manpower policies.

Industry structure and firm strategies as well as public policy have differed in the United States, Western Europe, and Japan in each of the three technological eras in the history of the semiconductor industry. In the 1950s—the transistor era—there were two types of transistor producers in the United States. The first were the established vertically integrated receiving-tube ("valve") producers who were also producers of radios and other electronic final goods. Such major firms included AT&T, RCA, Westinghouse, Sylvania, Raytheon, and GE. The second were new entrants who were either (1) not established producers of receiving tubes but were final equipment producers such as computer manufacturers or telecom equipment manufacturers (e.g., Motorola, Honeywell, and Sperry Rand) making many components in-house, or (2), merchant producers of semiconductors (e.g., Texas Instruments and Fairchild). Merchant producers became innovation leaders in ICs from the 1960s on. European and Japanese industry were quite different in that they were dominated through all three technological regimes by vertically integrated final goods, mainly consumer goods, makers, and to a lesser extent by telecom equipment makers, among the most important of which were Philips, Siemens, NEC, Hitachi, Toshiba, and Fujitsu.

If we categorize firms by strategies, aggressive major innovators were essentially confined to the United States till the late 1970s. Among these the stars were the new-entrant merchant producers such as Fairchild and Texas Instruments in the 1950s and 1960s, and Intel, National Semiconductor, and Mostek in the LSI era. However, there were also many firms that followed minor, or incremental, innovation strategies, entering the market as second sources for devices designed by major innovators and concentrating on incremental innovation to reduce costs. In Europe and Japan most firms remained incremental innovators in strategy, often producing under license from US firms.

Looking at the evolution of the industry from the demand side, the market for semiconductors, particularly for ICs, has been much larger in the United States than in Europe and Japan, and larger in the latter than in individual European countries. US demand in the 1950s and 1960s was largely from the military, computer, and industrial equipment segments and much less from consumer electronics. This biased demand toward ICs more than transistors in the 1960s, for which it created scale economies and acted as a major stimulus for innovation. In Europe, on the other hand, as in Japan, consumer electronics demand dominated until the 1970s, biasing demand toward discrete devices. European demand even in the 1970s was dominated by consumers and, by then, telecom demand due to lower levels of computerization than in the United States and Japan. Japan's computerization proceeded far more rapidly than in Europe. By the 1970s, Japan's semiconductor demand came increasingly from computers and sophisticated consumer electronics. It was this shift in the composition of demand due to the rapid growth of chip-intensive equipment segments that stimulated IC innovation and competitiveness in Japan, especially in CMOS memory chips. For this reason, Japan overtook Europe and began to challenge the United States in ICs from the late 1970s on. In addition, by the late 1970s vertical integration in Japanese firms and their conglomerate structure began to become an advantage in surviving semiconductor recessions and making lumpy investments in an increasingly capital-intensive technology.

Public policies have also had differing histories in the three major markets. In the United States, public procurement and R&D support by the defense and space programs provided consistent support in the 1950s and 1960s. However, from the late 1960s on, public policy was not as consistent and supportive as it had been earlier, especially as the civilian computer, telecom, and industrial electronic markets set the pace for the rate and direction of technological change. In contrast to the United States, European policies were weak and inconsistent until around the early 1980s, when restructuring of the vertically integrated firms took place in response to lagging competitiveness. Japanese policies were like those of Europe in the 1950s and 1960s, but in the 1970s MITI systematically and successfully targeted the IC industry, especially CMOS memory chips, for support. This was crucial to the emergence of a competitive Japanese semiconductor industry.[5]

HOW DID THESE TECHNOLOGICAL CHANGES IN SEMICONDUCTORS AFFECT THE EVOLVING MARKET STRUCTURES IN ELECTRONIC FINAL GOODS INDUSTRIES?

Miniaturization and Modularization: The Lowering of Techno-Economic Entry Barriers in the Electronic Equipment Industry

The advent of the IC, and later the microprocessor, revolutionized electronic equipment industries, unleashing two important technological trends at the equipment level: *miniaturization* and *modularization*. These two trends, in conjunction with the falling prices that they caused, reshaped market structures in the international computer and telecommunications industries and led to a convergence between the two. This created opportunities for entry into computers, telecom, and control systems for newcomers, including those from developing countries.

Miniaturization took place fundamentally at the chip level as function densities increased exponentially over the IC and microprocessor eras. More processing power and memory capacity than ever before could be packed onto a shrinking area of silicon wafer. It became possible to miniaturize computers and other electronic equipment while greatly improving their performance characteristics as price–performance ratios improved as costs per function plummeted. This represented a lowering of entry barriers for new firms aiming at the emerging low end of the computer market, including those from developing countries. This trend was also visible in other digital electronic equipment.

This increasing function density at the chip level was reflected in the plummeting costs per megabit of main memory, from $6,000 in 1960 to $90 in 1976, to single digits in 1987![6] The 16-Kbit DRAM was developed in 1978 and led to still more powerful and smaller-sized minicomputers and the emergence of the microcomputer. The 64K DRAM appeared in 1981 and the 256K DRAM in 1984, both in microcomputers. In a later section on developments since the late 1980s more recent developments are presented.

Similar progress was observed in logic chips, the chips at the heart of the central processing unit (CPU) of a computer, which actually performs the logical operations. The speed of a logic chip, which from 1971 on included microprocessors, is determined by the word length in bits of the data that it can process at a time. Intel developed the 8-bit microprocessor family in 1972, the 8008 series; the 16-bit microprocessor, the Intel 8086 series appeared in 1978 and the 32-bit chip in 1984. These led to even more powerful minis, the emergence of micros as powerful as old minis and older mainframes, and eventually the 32-bit chip-based superminis and supermicros of the mid-1980s and the 64-bit chips of the mid-1990s.

Parallel to these developments were slower but greatly cost-reducing innovations in peripherals which today make up the bulk of the cost of a computer system. Floppy disks, a 1971 IBM innovation, hard disks, printers, monitors, and modems improved in price–performance characteristics. The result was a series of smaller, cheaper, and more powerful minis, and from 1980 on, micros. This, in turn, resulted in mainframes and even minis being replaced by networked micros. Such flexible, reconfigurable systems revolutionized office automation and industrial process control.

The semiconductor revolution also changed the character of the telecom equipment industry, leading to an ongoing convergence of computer and communications technologies, both being increasingly based on common basic chips. The key technological shifts have been the move from analog to digital transmission and from electromechanical (crossbar) switching equipment (exchanges) to space/frequency division multiplexed (S/FDM) electronic exchanges—that is, analog electronic exchanges. Time division multiplexed (TDM) exchanges—that is, digital electronic exchanges—arrived later. They increased tremendously the capacity of exchanges in handling large volumes of signals (e.g., telephone calls). Main exchanges (public switching equipment) had been the major bottleneck impeding the expansion of the telephone network and their backwardness responsible for wrong numbers or inability to make connections. These innovations were facilitated by the development of coaxial cable, ultra-high-frequency, microwave, satellite, and now fiber-optic cable transmission technologies.

Any telecommunication network consists of three categories of equipment: (1) subscriber-end equipment—for example, telephones, teleprinters, telex terminals, and private automatic branch exchanges (PABXs); (2) transmission equipment—for example, transmitters, cables, and relays; (3) public switching equipment—that is, main exchanges. The latter are at the heart of the system and perform the function of switching electrical signals (e.g., telephone calls) from incoming lines to outgoing lines.

The earliest switching equipment was the strowger telephone exchange based on innovations of the late nineteenth century. The next generation were crossbar exchanges developed in the 1920s, which progressively replaced strowger over the postwar period but remained in place in many developing countries. Both these are electromechanical technologies as against fully electronic exchanges developed since the 1960s. As the number of telephones and other terminals grows, the number of connections to be made grows geometrically. The efficient, swift, and reliable switching of greatly increased numbers of incoming lines to outgoing lines began to be beyond the capabilities of electromechanical exchanges, and they became bottlenecks.

Electronic exchanges based on semiconductor technology were first developed at AT&T's Bell Labs in 1965 and were based on transistors. They used a multiplexing technique known as space or frequency division multiplexing

(S/FDM) or switching (S/FDS), which made possible the stored program control (SPC) switch. Stored programs in the exchange equipment made possible more efficient switching. These electronic exchanges could switch incoming analog signals only. At about the same time the concept of pulse code modulation (PCM)—that is, converting analog to digital signals—was used to pioneer digital transmission. The development of these early electronic exchanges and of digital transmission were combined to develop a more advanced electronic exchange based on the time division multiplexing (TDM) technique by Bell Labs in 1976. These exchanges were based on the then latest semiconductor technology—that is, very-high-speed bipolar digital ICs. This fully digital switching equipment could switch digital signals and was immensely more powerful than the still existing electromechanical exchanges and FDM exchanges. It was also possible to offer far more and newer telecom services with these fully digital exchanges, including the data-transfer capability necessary for networking computers. The overall trend had been toward ever smaller, more powerful, accurate, reliable and cheaper (per line) switching equipment and broader-bandwidth transmission capacity capable of handling ever-larger volumes of electrical signals (in bits/second) carrying voice, pictures, text and other information, enabling long-distance data transfer and national and international computer networking.

Miniaturization at the chip and equipment levels reduced size and costs while increasing performance and hence lowered barriers to access for potential users and barriers to entry for potential new manufacturers. Another technological trend—modularization—lowered these barriers further by increasing technological divisibility and thereby reduced the technological complexity for a potential entrant. Electronic equipment increasingly became modular in construction. They were composed of discrete modules representing different parts of the total system, with their circuitry designed around fewer and more powerful chips for the same functions. The systems were flexible and could be altered and reconfigured for different end-user needs by adding/dropping/changing the different modules of the whole system without having to redesign the circuitry of the module itself.

This trend opened the doors to entry into electronics manufacturing through the systems integration or systems engineering route. New manufacturers would buy components and modular subassemblies off the shelf and design the systems architecture and the operating systems software without necessarily having to design and manufacture chips or peripherals, in contrast to the vertical integration required by electromechanical technologies. Thus entry into low-end equipment such as personal computers (PCs) was made possible. Software has also become modular, affecting operating systems, utilities, and applications. A host of new entrants to computer manufacture worldwide arrived by this route.

The lowering of technological and minimum-investment entry barriers changed the market structure of the international computer, telecom, and other electronic equipment industries. Until the 1960s, the international computer industry was overwhelmingly dominated by the American mainframe manufacturers—IBM and the BUNCH (Burroughs, Univac, National Cash Register, Control Data, and Honeywell). From the late 1960s on, several new companies grew faster than the established ones with the growth of the minicomputer industry and emerged as giants, particularly DEC, Data General, and Hewlett Packard. DEC became the world's number two computer maker by the early 1980s. DEC's PDP series was a market leader in minis, followed by its VAX series.

A major deconcentration of the computer industry market structure took place in the minicomputer era of the 1970s, further accentuated in the microcomputer era of the 1980s, and only began to be reversed since the late 1980s with the commodification of PC production. This deconcentration took place not only with the growth of small and new US firms but also with the growth of Japanese and other foreign firms in this segment.

In the micro era, the pioneers were newcomers such as Apple, Radio Shack, and Commodore, which were later joined by the established mainframe and minicomputer vendors (e.g. IBM and DEC), as well as a host of new entrants. Most of the new entrants bought their parts and components off-the-shelf and competed in producing the most appropriate architecture for various user segments. The number of components and peripheral makers in this era of miniaturization and modularization also grew, resulting in deconcentration of market structures.

Most of these new mini- and micro manufacturers did not have extensive international networks of manufacturing subsidiaries and sales outlets as did IBM and the established large-system makers. This deconcentration of supply and the resulting increased competition among US, Japanese, and other firms for world (including developing-country) markets, in which they did not have a manufacturing and installed base presence, combined with the increasing availability of cheaper, more powerful, modular components and parts, increased the opportunities for technology acquisition by licensing. It also lowered the investment and know-how thresholds necessary for successful absorption of imported technology by developing-country entrants. This was because the new firms competing for world markets in a crowded field and lacking a presence in Third World markets were more amenable to licensing out their technology and to joint ventures with local firms than established giants such as IBM.

In process control, electronic instrumentation, and telecom equipment the picture was pretty much the same. The telecom market also experienced a drastic liberalization including deconcentration of suppliers and increased competition in the digitalization era of the 1980s generated ultimately by microelectronics-based technological change.

There has been a technology-pushed global wave of deregulation, competition, and falling prices for equipment and communication services, combined with increased power and versatility. The slow-changing system of national state-owned postal, telegraph, and telephone (PTT) monopolies have been progressively deregulated under the pressure of users whose needs, in terms of advanced communication services at reduced costs, the PTTs are unable to meet in comparison to the new private communication service firms. The most dramatic shift has been the breakup in 1984 of AT&T, America's equivalent to most of the rest of the world's PTT monopolies. New communication networks, services, and products were offered by firms such as GTE, MCI, and Rolm, as well as by European and Japanese competitors such as Plessey, Siemens, CGE, Jeumont-Schneider, Ericsson, NEC, and OKI. There has been a wave of mergers, acquisitions, and alliances.

The emergence of a deconcentrated market structure from the mid-1980s (which is showing signs of a tendency toward reconcentration for reasons discussed above for computers) opened up opportunities for entry into smaller, modular low-end telecom equipment for developing-country newcomers, as occurred with computers.

Technological and Market Shifts in World Electronics in the 1990s: The Closing of Old Windows and the Opening of New Windows of Opportunity?

Since the late 1980s, there have been fundamental technological and associated market-structural shifts in the world electronics industry that, by the mid-1990s, have again reshaped the international opportunity structure that developing countries face and in which existing firms have to survive. In this section, we briefly outline these shifts and focus especially on how technology is increasingly blurring the distinctions between the various subdivisions of the electronics industry, and the policy implications of such developments.[7]

Since the evolution of the industry has been predominantly technology-driven at both product and process levels, and only secondarily demand-driven or policy-driven (principally, deregulation-driven), we begin with the semiconductor industry, the prime driving force of technological change. Semiconductors have shown exponential increases in function density, speed, memory capacity, processing power, and versatility with corresponding improvements in price–performance ratios. The era of VLSI is giving way to the era of ultra-large-scale integration (ULSI), with densities of the order of tens to hundreds of millions of components on a chip, compared to hundreds of thousands in the mid-1980s. The memory capacity of DRAMs had gone up to 256MB by 1994, although 16MB still dominated the market, with 64MB scheduled to take over in the later part of the year. This is a thousandfold increase over the 256KB capacity that was introduced in 1984 and had dominated the market by the late 1980s. Microprocessors are shifting from 32-bit to 64-bit, thus greatly boosting processing speeds, the latest 32-bit microprocessors giving speeds of up to 10 million instructions/second (MIPS)—that of mainframe computers of earlier generations. Optoelectronic devices, including optoelectronic ICs (first developed by Bell Labs in 1990), are also emerging fast, alongside the increasing use of fiber-optic transmission and the integration of computers and communications.

All this has been made possible by linewidths shrinking to submicron levels, down to 0.35 microns for 64MB DRAMs by 1993, wafer sizes increasing to 8 inches for 64MB DRAMs, access times dropping and clock speeds increasing sharply. The same trends are in evidence for other types of ICs, both standard and ASICs. By 1992, the price of 1MB DRAMs had fallen to $10—that is, 1 millicent per bit of memory. These trends continue at a breakneck pace, guaranteeing the obsolescence of any technical update in a book on the industry by the time of publication.

This tremendous increase in processing power and memory capacity and the associated decrease in unit costs have made possible a tremendous increase in the power of microelectronics-based equipment and systems at the hardware level along with a reduction in size and costs. The latter has also been helped along by the spread of surface-mount technology compared to insertion of wired components into printed-circuit boards, a key development in component assembly technology. It has also made possible a tremendous increase in the versatility of equipment by increasing divisibility, hence modularity, and has thereby triggered revolutions in systems architectures.

An integrated "information technology" is rapidly gathering momentum due to these developments. This consists of the convergence of not only computer hardware and software with telecommunications but also with broadcasting, consumer electronics, industrial electronics, and control instrumentation. Within these trends, several hybrid technologies have emerged and are themselves evolving rapidly. The main trends in the major segments of electronics are presented below.

The major trends in the international computer industry have been the following:

(1) *The emergence of extremely powerful PCs and workstations* that can rival the performance of earlier generation minis—even mainframes—leading to a relative decline in the mainframe market in tandem with the trend toward downsizing of machines and networking. In 1985, 32-bit microprocessors were introduced; in 1993, Intel's Pentium heralded the advent of 64-bit microprocessor-based PCs.

(2) At the high end, *increasingly powerful supercomputers* have emerged, moving from the hundreds-of-megaflops performance range of the mid-1980s to the gigaflops range by the early 1990s and soon to the teraflops range. Neural and optical computers are future possibilities. But the major development at the high end has been the emergence of massively parallel systems employing nonconventional (nonsequential) architectures built on specialized or even standard microprocessors such as the Intel 80486. These are squeezing traditional Cray-type supercomputers and mainframes from the high end, while increasingly powerful PCs and workstation networks have been squeezing them from the low end.

(3) *Mass data-storage and retrieval systems*—a requirement for the effectiveness of newer and more powerful computers and networks—are being devel-

oped to remove bottlenecks that will negate the improvements in processing power.

(4) The movement toward *open systems*, or, in other words, *interconnectivity* or *interoperability*, makes possible interconnection of hitherto incompatible hardware and software of all kinds, both computers as well as other types of electronic equipment including telecom, industrial, and consumer electronic equipment. The precursors to this were the explosion of PC clones based on IBM compatibility and the emergence of an independent software industry in the 1980s, all made possible, ultimately, by the modularity of computer architectures. The trend toward open systems has been made possible not only by such modularity based on developments at the chip and peripheral levels, but, most importantly, by the trends in the software industry toward open operating systems. This makes it possible to run hitherto incompatible software on any open operating systems and to run such operating systems on all kinds of computers. A whole new range of possibilities have opened up for systems integration. These trends are reinforcing the revolution in computer network architectures that are associated with downsizing and distributed data processing. Open systems interconnection standards are still evolving.

(5) Yet another major trend is the explosive growth of computer networks. Between 1989 and 1993, the proportion of computers in the United States connected in networks rose from under 10% to over 60%, and of those in local area networks (LANs) to 52%. In Europe LAN-connected personal computers were estimated to be 33% of the total PC population, while in Japan it was still under 9%. There are over 100,000 computer bulletin boards in the United States; over 2.5 million computers and a multiple of that number of users are connected to the Internet, the "network of networks" that facilitates global data interchange to which as many as 18,000 networks are linked in the United States alone.[8]

Trends in telecommunications, broadcasting, and networking, leading to their convergence with computers, can be summarized as follows. The principal innovations are in high-speed digital switching and in digital transmission technologies which, taken together, make available vastly increased bandwidths both in cable and over-the-air transmission and facilitate huge increases in high-speed data transfer of all kinds.

In digital switching the most significant recent development is code division multiple access (CDMA) technology, which is taking over from time division multiple access (TDMA). While TDMA has managed to squeeze up to seven times as many users into the same slice of spectrum as analog technology, CDMA can squeeze up to twenty times as many, thus contributing to the explosive spread of over-the-air tele- and datacommunications. In addition, futuristic technologies such as photonic switching are being developed.

The other revolutionary development in telecom is in digital data-compression and data-transmission technologies, which make possible digital transmission not just of sound (telephony) but all other kinds of signals, even very-wide-bandwidth-requiring forms such as TV and video images. All three major transmission technologies—coaxial cable, satellite, and fiber-optic cable—have been making rapid advances, especially the latter two. Fiber-optic cables have contributed to an explosion of hitherto scarce bandwidth.

New innovations are making available the microwave and possibly, in the future, the millimeter wave regions of the electromagnetic spectrum, contributing to an explosion of bandwidth over the air to match that over wire. Cellular telephony has shown explosive growth, with over 8.4 million cellular telephones in the United States by 1993 in one decade of existence. Flowing from this, cellular data transmission, using cellular digital-packet data technology in which such packets are transmitted in high-speed bursts between calls, is making rapid progress. All in all, digital transmission systems of bit rates of 2.5 gigabits/second have been installed and next-generation systems of 10 gigabits/second are in the pipeline.

All this, taken together with digital switching innovations, is promoting the wireless/cordless telephony and datacommunications revolutions and giving a fillip to the market for all kinds of cordless equipment and systems and, in the process, also promoting distributed dataprocessing in decentralized network architectures, as these networks need not necessarily be wired. These developments, in turn, make possible vastly more sophisticated applications of information technology in manufacturing and office automation, including increasingly "intelligent" systems with rule-based decision-making capability for myriad applications.

These technological developments are revolutionizing the centralized, top-down architectures of the three traditional networks—the telephone network, the TV broadcasting network and computer networks. Traditionally, all three networks were designed top-down to address the problem of lack of power at the subscriber (terminal or peripheral) end and the related fact of scarcity of bandwidth (spectrum) that made two-way communication difficult except for telephony where a call requires only 4 Kilohertz of bandwidth. Data other than voice, and especially TV signals, require large quantities of bandwidth and very high-capacity switches, only possible with recent digital switching innovations. Therefore, the networks were centered around powerful mainframe computers, centralized TV transmitters, and main exchange switches, respectively, with the receiving (user) end having only (comparatively) feeble computer terminals, TV receivers, and telephones.

The revolutions in digital switching capability and the bandwidth explosion due to developments in digital compression and transmission technologies make possible fully interactive networks, "empowering" the receiving end while undermining the importance of centralization. This has occurred, first of all, in computer networks, moving away from mainframes toward decentralized client–server architectures built around vastly more powerful PCs and workstations. This shift is driven by the need for organizational flexibility and economy, since client—server architectures, usually based on open standards and interconnectivity, give access to data and information at the point where it is needed.

Telephone and TV networks are following the trajectory of computer networks by empowering the receiving end by vastly enhancing the power of the telephone and the TV set. The latter are increasingly becoming multipurpose interactive terminals that are capable of both receiving and sending a multiplicity of signals (sound, text, graphics, images, television, video)—thus presaging the eventual emergence of the hand-held personal digital assistant capable of receiving and transmitting every kind of signal from anywhere to anywhere through global fiber-optic/satellite/cellular networks.

Pulling the threads together, the developments since the late 1980s have led to the following trends in the world electronics industry: (1) microminiaturization at the chip level has led to smaller, lighter, portable, yet more powerful and versatile electronic equipment of all kinds. (2) Electronic equipment of all kinds is increasingly becoming interconnectable/interoperable: equipment of one make can interface with that of another, and systems architectures and networks can be built by the integration of hardware and software of diverse types and makes. The key to this is the trend in software toward open operating systems. This amounts to a tremendous extension of the modularity that made possible IBM-compatible microcomputer clones in the 1980s. Furthermore, the boundary between electronic and nonelectronic machinery is getting blurred as the latter become electronified. (3) The top-down systems architectures of telephone, broadcasting, and computer networks are being decentralized by the empowerment of the receiving end by increasingly powerful, versatile, and hybrid equipment in increasingly hybrid networks capable of two-way interactive transmission of all kinds of signals—voice, text, graphics, video, and so forth. (4) Wireless/cordless/over-the-air tele- and data-communications are revolutionizing telecom, datacom, broadcasting, and computer networks, making possible a host of new products and applications. (5) All these developments, taken together, are promoting the emergence of intelligent (rule-based inference-capable) computers, sensors, machines, databases, and networks.

The implications of these trends for evolving market structures in the world electronics industry and for the opportunity structure and policy options facing developing countries are, very broadly, as follows since the late 1980s. Entry into the world electronics industry through the windows of opportunity that had opened up in the 1970s and early 1980s is no longer a viable strategy. This is due to technological and economic forces at work that tended to raise entry barriers once again in those segments. For example, the automation of semiconductor assembly meant the recentralization of such operations, closing such a route for countries other than those such as Korea and Taiwan, which had successfully developed indigenous design and fabrication capabilities; the development of scale economies in PC cloning and other OEM-type mass production, raising financial and marketing barriers to entry; the development of software automation, the tendency toward larger firms in the software industry due to high development and marketing costs; and the technological and skill barrier posed by the shift to fourth-generation languages (4gls). However, alongside these adverse developments, the recent trends potentially open new windows of opportunity for developing countries. The challenge for policy, then, is to be flexible enough to take advantage of these new opportunities and not get stuck in segments entered earlier in which the odds are increasingly stacked against success.

Before describing the evolution of opportunities consequent to the industry trajectory described, it should be noted that earlier strategies, such as OEM-type contract manufacturing on a mass- or batch-production basis are still possible by domestic firms and joint ventures or by relocation of plants from countries losing competitiveness. The only difference is that the relocators are likely to be NICs such as Taiwan and Korea losing competitiveness due to rising wage levels, and that the products relocated are likely to be their formerly successful specialties while they move on to higher value-added products.

However, to outline the restructuring of business triggered by these trends, basically what has been happening is the convergence of the computer, communications, consumer electronics, and entertainment industries as their products become merely different systems architectures built around the same types of components and are increasingly interconnectable. This is creating new markets for existing players in each of these subdivisions of the electronics complex of industries and has been behind the trend toward alliances, mergers, and acquisitions within and across existing subdivisions and spanning international borders and oceans, to equip oneself better with appropriate partners to exploit these new opportunities. In the process, new types of specialization and niche markets are opening up for both old and new players.

To get a head start in creating the digital multipurpose networked (possibly wireless) computer–communications–consumer-electronic–industrial-automation equipment of the future, computer companies are talking to consumer electronics companies, consumer electronics companies are talking to entertainment and information industry companies (especially cable television with its high bandwidth capability), telephone companies are talking to computer and consumer electronics companies, and all of them are talking to component and software companies. The idea is to bring together the processing and switching power of computer and telecom firms with the bandwidth capability and entertainment software of cable television and the entertainment industry, the manufacturing capabilities of consumer electronics firms, the basic technologies of component makers, and the open operating systems capabilities of software companies that are necessary to interoperate the new systems.

This inevitably leads them to strike international alliances to find the best partners, spread costs, and minimize risks, since the Japanese lead in consumer electronics, the Americans in computers, software and digital programming, both have strengths in semiconductors, and the Europeans have strengths in telecom and satellites, besides a vast market. This has led to former rivals getting together, very often so as to avoid the risk of getting stuck with a nonstarter product in a world where standards are still evolving and uncertainties prevail.

A corollary of these trends since the late 1980s has been the tendency to concentrate on what one does best and hive off less central activities and outsource inputs, including services, from those who do it best on a global scale. This has created a new set of opportunities for developing countries, the most important of which has been software exports, including the opportunity to upgrade to more complex and comprehensive software development. Low-end labor-intensive coding-type jobs are now increasingly threatened by software automation; survival lies in upgrading. The second has been to export more sophisticated hardware by mass or batch contract manufacturing on an OEM or even brand-name basis, including more powerful chips, computers, and telecom equipment. This option is open to those countries that have or can create a competitive manufacturing base by policy changes. The third option, which combines existing hardware and software capabilities, is systems integration or systems engineering on a much broader basis than earlier, in an increasingly technologically convergent and interoperable electronics world. This is potentially less threatened by automation as most systems development is customized, intensive in skilled labor, and offers a great variety of niches. It also dovetails with the spread of domestic applications of electronics, which would require, typically, customized solutions.

All these would require the striking of appropriate strategic alliances with global players in the field leveraging the manufacturing base, technological capabilities, and human resources developed in the earlier phases of entry into the world electronics industry. This would, in turn, again require appropriate policy shifts as well as policy flexibility at both industry and macroeconomic policy levels and, by implication, a shift in the ideology and strategic interests of the coalition of forces dominating state policy.

POLICY IMPLICATIONS FOR DEVELOPING COUNTRIES

Entry Possibilities and Strategic Options

International technological and economic trends in the electronics industry opened a "window of opportunity" for entry into electronics manufacturing for the more advanced developing countries, starting in the late 1960s. Entry into the consumer electronics industry became possible by assembling components bought off-the-shelf. This was a labor-intensive assembly industry, and an LDC firm could act as an OEM for a developed-country manufacturer or distributor, thus accessing developed markets. Alternatively, it could cater to a protected domestic market and hope to become internationally competitive later after gaining production experience in the domestic market, provided economies of scale could be reaped on the basis of domestic demand.

Entry into semiconductor component production was also possible along similar lines. From the late 1960s on, the major US semiconductor firms were shifting the labor-intensive assembly stage of semiconductor production to low-wage areas to reduce costs. It now became possible for developing countries to find a niche in semiconductor assembly in the hope of eventually integrating backward vertically into fabrication or even design, and forward into final testing, in joint ventures with MNCs if necessary. The 1970s was still the era of manual assembly operations, and relocation to low-wage areas was still economical for MNCs.

Semiconductor manufacturing possibilities were much more limited than consumer electronics assembly in that ISI was impossible without a semiconductor-consuming equipment manufacturing industry. Few if any developing countries had sufficient domestic demand for semiconductors. And assembly plants relocated by MNCs would be 100% owned by them as it was a crucial stage of a vertically integrated manufacturing operation in which absolute reliability of delivery was essential. Transfer of technology would be minimal, and there would be plenty of scope for transfer pricing. However, it was still an avenue into the semiconductor industry that generated employment and exports and held out the possibility of upgrading to automated assembly, testing, fabrication and design.

More significantly, from the 1970s on, a window of opportunity opened for entry into mini-, and later into microcomputer production and software exports, and still later into low-end telecom. The systems integration route, which emerged due to the lowering of economic and technological thresholds, became a viable path to the creation of an indigenous computer industry initially based on the domestic market. Brazil and India embarked vigorously on the creation of a domestic computer industry in the 1970s. Korea followed suit in the 1980s, but along OEM lines as a supplier to major world producers, as it had done earlier in consumer electronics.

Simultaneous Entry into Many Segments of Electronics: The Options

Entry into more than one segment of the electronics industry simultaneously was also possible in a number of different ways requiring different policy combinations. One option was to create an export-oriented local manufacturing base in semiconductor assembly and consumer electronics on a foreign-ownership or joint-venture basis, within a liberal trading and foreign investment regime, while depending on imported components and parts. There would be few forward or backward linkages between the consumer and component segments in such a strategy, the latter being aimed at the world market in a largely intra-firm operation and not dependent on the former to provide it sufficient demand. After a period when the demand for chips justified it, vertical integration into chip fabrication, final testing, and design could be attempted. So could upgrading of the equipment industry into more sophisticated and complex manufactures such as computers and peripherals, and low-end telecom equipment on an ISI or an OEM export-oriented basis.

Alternatively, simultaneous entry into the consumer, component, and even the more complex computer and industrial electronics segments could be attempted on a mainly ISI basis. The equipment segments in such a strategy would follow the systems-engineering route, and the component segment would perforce be mainly an assembly industry at first, both dependent on key imported inputs. Such a strategy, if it was to have a chance of success, required massive complementary investments in the expansion of broadcasting, telecommunications, and computerization of government, the financial sector, and the economy in general to provide demand for the equipment and, indirectly, component industries, given very small domestic markets.

Even then such an import-substituting industry, especially components, would run very high risks of inefficiency. In a predominantly ISI strategy, specialization in a few standard products as well as phasing of vertical integration into more advanced products, depending on the expected market size for the product concerned, would have to be carefully planned. Specialization, standardization, and phasing would be the most critical choices to be made. Despite its difficulty, an ISI strategy could conceivably be justified on grounds of protection of technological learning for the creation of future comparative advantage, that is, in dynamic-efficiency terms. It also could be required by, or forced on the country due to, national security considerations, or due to lack of access to developed markets or lack of MNC investor confidence in the country for political reasons or policy-induced investment climate reasons.

Finally, a mixed strategy of ISI in some segments such as computers or telecom via the systems integration route could conceivably be combined with an export-oriented MNC-led entry into component or consumer electronics assembly.

Strategic Requirements for Entry into Electronics

All these entry strategies into the electronics industry require, first of all, access to technology and capital and, second, access to markets—international, protected domestic, or both. Both these requirements force politically and economically difficult choices with regard to foreign ownership, import of technology, infant industry protection, and public investment. An export-oriented MNC-led strategy creating a base only in the simpler labor-intensive consumer and component assembly segments implies a dependence on foreign capital, technology, and market fluctuations. On the other hand, an in-depth ISI approach would be threatened by inefficiency due to the limited markets of developing countries and lack of economies of scale. Such inefficiency would ultimately accentuate dependency, thus defeating the objectives of the strategy.

The planning and implementation of any of these strategies—whether MNC-led semiconductor and consumer electronics assembly, OEM-type export of assembled consumer electronics on a domestic firm or joint-venture basis, or a mixed ISI strategy for simultaneous entry or a thoroughgoing ISI strategy—would run into political constraints and would require the capacity to surmount them. An MNC-led or OEM-type export-oriented assembly strategy in semiconductors or consumer electronics would require the capacity to manage relations with foreign capital in such a way that technology transfer and domestic firm capabilities are built up so that upgrading to more complex manufacture is possible.

A mixed ISI strategy of simultaneous entry into assembly-type export production as above combined with domestic market-oriented equipment production would require not only this but also the capacity to make appropriate choices of entry, specialization, and phasing, critical levels of investment, the necessary complementary investments, and appropriate choices between public and private sectors. A predominantly ISI strategy would require these capacities in even greater measure since achieving the vital economies of scale would be both more difficult and more critical. Essentially, in all cases the state will need to be able to impose its priorities over the short-term interests of different sections of domestic and foreign capital and other organized groups, even those within the state apparatus.

As for the policy implications of world technological and market trends since the late 1980s, the principal implication is that developing countries have to have not only the autonomy but also policy flexibility and interagency coordination for a continual restructuring of their industries, policy frameworks, and policy institutions. This, in turn, would require complementary policies of appropriate openness to international firms combined with comprehensively supportive domestic policies on both supply and demand sides to buttress the capabilities and bargaining power of domestic firms in their strategic negotiations. It would also involve the willingness to make the inevitably painful decisions to phase out obsolescing activities, steer existing firms toward more competitive lines of production, and support the rationalization, merger, or exit of losing firms.

This is only a brief summary of the requirements of the various strategic options open to developing-country entrants. Korea followed a neatly phased strategy starting with MNC-led semiconductor assembly and ending in domestic-firm export of sophisticated products. Brazil and India followed predominantly ISI strategies, moving toward liberalization only in the later 1980s and 1990s. Why was Korea the most successful by the usual indicators and India the least, even though the latter embarked on advanced products much earlier and had a head start in important respects? What are the implications for the political economy of competitiveness and for the theorizing on the strategic capacity to develop new industries? These are the questions we try to answer in the following chapters.

NOTES

1. We follow the classification used in India, our principal case, to facilitate comparisons later in the text.

2. We use industrial electronics or industrial equipment henceforth to mean electronic equipment other than consumer electronic equipment—that is, encompassing categories (2) to (5).

3. In this section we are heavily indebted to Franco Malerba, *The Semiconductor Industry: The Economics of Rapid Growth and Decline* (Madison, WI: University of Wisconsin Press, 1985); United Nations Center for Transnational Corporations, *Transnational Corporations in the International Semiconductor Industry* (New York: United Nations, 1983); Robert K. Wilson, Peter K. Ashton, and Thomas P. Egan, *Innovation, Competition and Government Policy in the Semiconductor Industry* (Lexington, MA: Lexington Books, 1980); David C. Mowery, "Innovation, Market Structure and Government Policy in the American Semiconductor Electronics Industry: A Survey," *Research Policy*, 12, no. 4 (August 1983); B. Bowonder and K. Vinod Reddy, "Microelectronics: State of the Art," *Electronics: Information and Planning*, 19, no. 10 (1992): 511–544; numerous articles in the general and industry press, especially *Electronics* and *Electronics Business*.

4. Extremely high-speed and radiation-hardened (resistant) gallium arsenide (GaAs) chips are used for specialized equipment, especially for defense/space applications; optical (photonic) chips and biochips are possibilities but are still largely at the R&D stage at this time.

5. See Marie Anchordoguy, *Computers, Inc.: Japan's Challenge to IBM* (Cambridge, MA: Harvard University Press, 1989).

6. The following section draws heavily on Joseph M. Grieco, *Between Dependency and Autonomy: India's Experience with the International Computer Industry* (Berkeley and Los Angeles, CA: University of California Press, 1984), pp. 53–70; B. Bowonder and K. Vinod Reddy, "Microelectronics"; B. Bowonder and T. Monish Singh, "Information Technology—State of the Art," *Electronics: Information and Planning*, 19, no. 8 (1992): 397–429; Michael Hobday, *Telecommunications in Developing Countries: The Challenge from Brazil* (London: Routledge, 1990), pp. 35–87; Christiano Antonelli, *The Diffusion of Advanced Telecommunications in Developing Countries* (Paris: OECD, 1991); Dieter Ernst and David O'Connor, *Competing in the Electronics Industry: The Experience of Newly Industrializing Economies* (Paris: OECD, 1992); Nancy S. Dorfman, *Innovation and Market Structure: Lessons from the Computer and Semiconductor Industries* (Cambridge, MA: Ballinger, 1987).

7. For the summary of world electronics trends in this section we draw heavily on the following sources: Bowonder and Reddy, "Microelectronics"; Bowonder and Singh, "Information Technology"; Antonelli, *The Diffusion*; Ernst and O'Connor, *Competing*; B. Bowonder and D. Poorna Chander Rao, "Microelectronics: State of the Art and Imperatives for India," *Electronics: Information and Planning*, 20, no. 8 (1993); and the electronics and business press.

8. See *Newsweek*, 6 June 1994, cover story, for all figures in this paragraph.

3

The Political Economy of Export-Led Electronics Strategy in Korea

ELECTRONICS MANUFACTURING IN KOREA

The Empirical Record in Brief: Three Phases and a Possible Fourth

Korea is the polar case of successful export-led growth in electronics led by domestic private firms and resulting in progressive vertical integration backward into more sophisticated products. Korean electronics manufacturing began with local assembly of radio receivers for the domestic market from imported components by domestic firms in 1959, followed by black-and-white (b&w) TV sets, stereos, and radio communication equipment. This initial ISI phase ended in the mid-1960s, after which there was a shift toward exports, following the economy's turn toward export orientation. From then on, the world market, specifically the US market, became the main source of demand for Korean electronics. From the late 1960s on, foreign firms, mainly American and Japanese, began to relocate labor-intensive component assembly in Korea and re-export the subassemblies to their home plants for final testing. By the late 1960s, MNCs accounted for a third of output and three-quarters of exports, overwhelmingly of components. However, Korea had a trade deficit in electronics because MNCs and joint-ventures imported all their raw materials.[1]

In 1969 the Korean government passed the Electronics Industry Promotion Law, and an eight-year electronics industry development plan was introduced and implemented, with a focus on export promotion. Korean electronics exports boomed in the 1970s against the backdrop of buoyant world electronics trade.

There were two identifiable streams of exports. The first was the export of component subassemblies, overwhelmingly dominated by wholly foreign-owned subsidiaries and 50/50 joint ventures, continuing the trend set in the late 1960s. A large part of this was done in the FTZs, where MNCs were given fiscal and financial incentives. The other stream was of consumer electronics, mainly of radio and TV receivers. Exports of these grew even faster than those of components and were dominated by domestic firms assembling imported parts according to the specifications of their foreign suppliers and re-exporting to the latter, who would market them abroad under their brand names.[2]

For developed-country consumer electronics vendors, this was a way of getting their manufacturing done cheaply by subcontracting to original equipment manufacturers (OEMs)—in this case Korean firms who had mastered efficient plant operation and who had a well-policed, low-wage labor force. It was by way of consumer electronics exports and, to a lesser extent, joint ventures for components assembly, that the giant Korean electronics firms, Samsung, Daewoo, and Goldstar (later Hyundai), all part of diversified conglomerates, grew into significant international competitors in the 1970s.

Five important points need to be noted here.

(1) The initial ISI phase of the 1960s was critical to the development of the manufacturing skills that enabled these firms to become the efficient consumer electronics and component assemblers of the 1970s. Indeed, ISI in consumer electronics parts and components continued in the 1970s after domestic demand from export production justified it.

(2) While Korea's share of world trade in both components and consumer electronics increased dramatically between 1970 and 1980, the share of components in total exports decreased in percentage from the eighties to the forties, while that of consumer electronics rose from the teens to the forties.

(3) As a result of the above trend, the share of domestic firms in exports rose steadily from 28% to 60% over 1971–83, while that of foreign firms fell from 60% to 31%.[3]

(4) The share of the technologically more complex industrial electronic equipment segments remained very small, rising over the decade to only 13% of production and 8% of exports by 1980 (Table 3.1).

(5) The electronics complex of industries in Korea exhibited very little of the synergy observed in developed country electronics. Until the late 1970s, there were few forward and backward linkages between the components and equipment segments. Each segment was oriented toward the world market.[4] The components segment imported its inputs and exported its products. So did the consumer electronics segment. There were very few linkages between the two. Foreign firms assembling components were entirely intrafirm operations.

The picture began to change in the late 1970s, as Korean electronics entered its third phase of moving into production and export of more advanced equipment such as personal computers, telecommunications equipment, and more advanced consumer products such as color televisions, VCRs, and so forth, as well as vertical integration backwards into such sophisticated areas as IC fabrication. This thrust was led from the early 1980s on by giant firms such as Goldstar, Samsung, Daewoo, and, later, Hyundai. As rising real wages and automation of electronics assembly processes began to erode their comparative advantage, these firms moved to create a comparative advantage in more advanced products. But as this would bring them into competition with the Japanese, in color televisions for example, it would therefore incur the threat of restriction of supplies of advanced ICs and other technology. Hence this strategy also required integrating backwards into IC design and fabrication. The technology for these thrusts has been acquired through licensing and joint ventures. Economies of scale were reaped by producing for Western firms on an OEM basis, as had been done earlier in consumer electronics—for example, producing terminals and monitors on an OEM basis for US computermakers utilizing skills acquired in TV production.

This long boom took Korean electronics production from a very small base of $56 million in 1968 to over $7.3 billion by 1985 and to $33 billion by 1991. Table 3.1 gives segment-wise output, export, and import figures from 1975 to 1991. Tables 4.5 and 5.3 to 5.6 contain comparisons with Brazil and India. Korea's output overtook India's by 1973 by national figures (Tables 3.1, 5.1) and Brazil's by 1982 or by 1987 using comparability-adjusted figures.[5] By 1985, Korea's $6.44 billion output was over three times India's, by comparability-adjusted figures, but about six times India's if measured at world prices. By 1990, using comparability-adjusted figures, Korea's output was nearly five times India's, despite faster Indian electronics growth from the early 1980s on (Table 5.3).[6] Korea's production of consumer electronics grew from a mere $33 million in 1971 to $2,411 million in 1985 and to $11,054 million by 1991. Its production of industrial electronics grew from $25 million in 1972 to $1,517 million by 1985 and to $7,104 million by 1991, and of components from $86 million in 1971 to $3,356 million by 1985 and to nearly $15 billion by 1991.

Korean electronics exports grew from $19 million in 1968 to $1,037 million in 1976, to $4,352 million in 1985 and to $19,334 million in 1991. Of this, consumer electronics exports grew from a mere $11 million in 1971 to $1,131 million in 1981, to $1,554 million by 1985, and to $6,054 million in 1991, reflecting the consumer electronics export boom of the 1970s and 1980s. Components exports rose from $42 million in 1970 to $975 million in 1981 and to $9,385 million in 1991. Exports of industrial electronics rose from just $29 million in 1975 to $133 million in 1981 and then jumped to $3,895 million in 1991, reflecting the jump in exports of industrial electronics in the 1980s. Their share rose to over 20% of exports, compared to just 6% in 1980.

One gets another and sharper picture of Korea's relative performance if we compare production and export figures of major products in numbers of units. Korean production of TV receivers rose from 1.225 million units in 1975 to 6.819 million in 1980, to 9.729 million in 1984, of which 4.614 million were color televisions, to nearly 16 million by 1991, almost entirely color televisions.[7] Brazil's TV production, in comparison, was about 1 million units in 1972, under

Table 3.1: Korean Electronics Production, Export, and Import by Broad Categories (US$ million)

Year	*Indicator*	*Indust*	*Consumer*	*Eqpt*	*Component*	*Total*
1975	Prod	94	270	364	496	860
1975	Export	35	199	234	384	618
1975	Import	72	61	133	312	445
1976	Prod	126	551	677	745	1,422
1976	Export	56	390	446	591	1,037
1976	Import	102	86	188	511	699
1977	Prod	185	679	864	894	1,758
1977	Export	103	436	539	568	1,107
1977	Import	122	76	198	640	838
1978	Prod	210	927	1,137	1,134	2,271
1978	Export	103	654	757	602	1,359
1978	Import	198	118	316	840	1,156
1979	Prod	320	1,374	1,694	1,586	3,280
1979	Export	163	925	1,088	695	1,783
1979	Import	262	219	481	687	1,168
1980	Prod	364	1,148	1,512	1,340	2,852
1980	Export	169	1,020	1,189	775	1,964
1980	Import	351	182	533	733	1,266
1981	Prod	494	1,574	2,068	1,724	3,792
1981	Export	173	1,198	1,371	799	2,170
1981	Import	493	192	685	846	1,531
1982	Prod	639	1,549	2,188	1,818	4,006
1982	Export	229	926	1,155	914	2,069
1982	Import	653	152	805	877	1,682
1983	Prod	945	2,189	3,134	2,426	5,560
1983	Export	298	1,164	1,462	1,242	2,704
1983	Import	796	207	1,003	1,226	2,229

Source: Electronics Industry Association of Korea (EIAK).
Note: Prod = Production; Indust = Industrial electronics; Eqpt = Total Consumer and Industrial Electronic Equipment.

2 million in 1976 (already falling behind Korea), and just over 3.5 million in 1980, remaining essentially stagnant over the decade to 3.27 million in 1991 (Table 4.5).[8] India reached the 3 million mark only in 1986 and peaked at 4.4 million in 1988, before four years of downturn (Table 5.8). Korean production of PCs, beginning in 1980, was over 1 million by 1984 and of terminals and monitors over 3 million. Compared to this, Brazil produced only 73,000 microcomput-

Table 3.1 (*continued*)

Year	*Indicator*	*Indust*	*Consumer*	*Eqpt*	*Component*	*Total*
1984	Prodn	1,213	2,426	3,639	3,531	7,170
1984	Export	687	1,794	2,481	1,970	4,451
1984	Import	937	269	1,206	1,879	3,085
1985	Prodn	1,518	2,411	3,929	3,356	7,285
1985	Export	905	1,860	2,765	1,825	4,590
1985	Import	957	249	1,206	1,735	2,941
1986	Prodn	2,117	4,814	6,931	5,164	12,095
1986	Export	1,449	3,058	4,507	2,742	7,249
1986	Import	1,246	362	1,608	2,735	4,343
1987	Prodn	3,707	6,977	10,684	6,754	17,438
1987	Export	2,353	4,904	7,257	3,876	11,133
1987	Import	1,348	517	1,865	3,932	5,797
1988	Prodn	4,573	9,211	13,784	9,747	23,531
1988	Export	3,225	6,436	9,661	6,070	15,731
1988	Import	2,287	622	2,909	5,219	8,128
1989	Prodn	6,097	10,247	16,344	12,291	28,635
1989	Export	3,485	5,948	9,433	7,131	16,564
1989	Import	2,573	670	3,243	5,683	8,926
1990	Prodn	6,345	10,141	16,486	12,432	28,918
1990	Export	3,481	5,727	9,208	8,016	17,224
1990	Import	3,068	724	3,792	6,057	9,849
1991	Prodn	7,104	11,054	18,158	14,946	33,104
1991	Export	3,895	6,054	9,949	9,385	19,334
1991	Import	3,530	743	4,273	6,973	11,246

EIAK figures include electrical appliances and are therefore more inclusive and larger than comparative figures based on Elsevier Advanced Technology, Oxford, UK, figures in later tables; however, the difference is of a not very large and fairly consistent order of magnitude, and more importantly, the trends revealed are consistent. All important trends such as sectoral growth rates of output, exports, imports, the trade balance and export orientation of production, the sectoral composition of production, exports, and imports can be computed from the above basic table.

ers in 1984 and India 10,000 in 1985.[9] Korean production of telephones grew from 2.3 million in 1982 to over 10 million by 1990, while Brazil's production of 1.13 million in 1983 never crossed 1.8 million and declined to almost 1 million by 1991. Indian production grew from 729,000 in 1983 to a peak of 1.62 million in 1990. The same is true for electron tubes (especially TV picture tubes), discrete and IC semiconductors and so forth.

The success of Korea's shift into more advanced consumer electronics, industrial electronics, and components is illustrated by the shift in the composition of its exports. Between 1975 and 1985, exports of what had been the bread-and-butter products of the 1970s export boom peaked and leveled off. In contrast to these products, exports of color televisions, monitors, ICs fabricated in Korea, and videotape recorders rose. The share of industrial electronic equipment rose to over 20% of exports by 1990; by comparability-adjusted figures, computers, telecom, and office equipment alone accounted for 20% by that time. And in components and consumer electronics, increasingly sophisticated products, such as advanced memory chips, came to dominate exports.

The significant changes in this third phase of the 1980s are the following:

(1) A substantial domestic market for electronics (larger than India's) had emerged by the early 1980s, due to rising per capita incomes and relatively egalitarian income distribution. This created a mass market for color televisions and even 8-bit home computers by the mid-1980s. In fact, by this time Korea's domestic market for electronics had become larger than India's. By 1989, its domestic market even for consumer electronics became larger than India's, even though Korea has one twentieth of the population of India. And by 1990, its market was $14 billion, compared to India's approximately $5 billion.

(2) The share of domestic firms in output in all segments and exports continued to rise from the late 1970s on. Particularly noteworthy is the rising share of domestic firms in the more complex industrial electronics segments, and in components.

(3) The exports of advanced consumer products and also of industrial equipment and components (including ICs) grew rapidly. Korea emerged as an exporter of such sophisticated products as the 256K DRAM chip as early as 1984, when Samsung became the first non-US and non-Japanese company to produce it.

(4) Electronics emerged as the leading export industry by 1988, ahead of textiles, and there had been a generally growing trade surplus in electronics every year since 1983.

(5) Significant R&D expenditures were made for the first time by major firms supported by R&D in government-supported electronics research institutes. The three major institutes were the Korea Electrotechnology and Telecommunications Research Institute (KETRI), Korea Institute of Electronics Technology (KIET), and Korea Advanced Institute of Science and Technology (KAIST). The first concentrated on telecommunications, the second on ICs and computers, and the third on basic research. In 1984, KETRI and KIET were merged to form the Electronics and Telecommunications Research Institute (ETRI), signifying a recognition of technological convergence. ETRI played a major role in technology development for the semiconductor, computer, and telecommunication expansion of the 1980s in public–private collaboration with the major *chaebol*s.

(6) There has been an increasing offtake of components by domestic industry, and for the first time significant synergetic intersegmental linkages, often intrafirm, have emerged in the electronics complex. The proportion of compo-

nents exported fell from 90% in 1971 to remain roughly in the three-fifths to three-quarters range by the late 1980s and early 1990s.

(7) There has been a gradual shift in the role of the state in the course of the 1980s from the earlier directive, strategy-led development role to a still critical facilitative or "big followership" role, letting the big firms lead, along with domestic deregulation and complementary infrastructural investment paralleling the post-1970s role of Japan's Ministry of International Trade and Industry (MITI).

The third phase of the 1980s seems to be giving way to a possible fourth phase since the end of the decade, characterized by the following features:

(1) As Korean wages and the won rose in the late 1980s and later entrants like Thailand, Malaysia, Philippines, Indonesia, China, and Mexico have begun to erode Korean competitiveness in labor-intensive industries, Korean industry is attempting to move up into newer technology products and compete with the United States and Japan. This tendency is accelerated by the growing openness and foreign penetration of the Korean market.

(2) This requires massive investments in new facilities and R&D and, hence, inward FDI for strategic tie-ups for technology, finance, and markets. Korea is also undertaking outward FDI for relocating labor-intensive electronics, mainly to Southeast Asia and China, but increasingly also to developed markets in consumer electronics.

(3) The world recession of the early 1990s and trade pressures from the United States and Europe have accelerated the trend to upgrade to higher value-added and independent brand name exports as against the present OEM pattern. Consumer electronics companies have been hit hardest. There is a massive ongoing effort to reposition for the rest of the 1990s by catching up technologically, especially after the conclusion of the Uruguay Round of GATT, with its implications for market openness and intellectual property protection.

(4) In response to these pressures, a number of major long-term initiatives have been launched with state support, with a focus on developing critical basic technologies. Korean companies are also increasing their FDI in the United States, often with the intention of acquiring high-tech firms.

(5) The role of the state is changing, along the same lines as in the 1980s—that is, from "big leadership" of the market to "big followership," providing infrastructural and R&D support rather than selective support of particular firms through directed credit and administrative guidance, along with liberalizing the Korean market for imports, technology licensing, and FDI.

The Growth of the *Chaebol* in Korean Electronics

No account of Korean electronics development would be complete without an account of the growth of the giant Korean electronics firms and the conglomerates of which they are a part, over the period from the late 1960s to today. For it was these large conglomerates [*chaebol*] and the industrial structure centered around them that made possible the rapid growth of Korean electronics exports, especially in the third phase of the 1980s.

The three firms that have dominated the Korean electronics industry are Samsung, Daewoo (after it acquired Taihan in 1981) and Goldstar. In the 1980s, these were joined by Hyundai. These firms are all part of conglomerates that are listed in *Fortune*'s list of 500 largest industrial firms outside the United States and, in recent years, in the *Fortune* Global 500 list. In 1980, the Hyundai group (petrochemicals, electronics, and appliances) ranked 72nd with sales of $5.54 billion, the Lucky (Goldstar) group (petrochemicals, electronics, and appliances) ranked 101st with sales of $4.45 billion, and Samsung group (electronics and appliances) ranked 125th with sales of $3.8 billion among the 500 largest non-US nonfinancial corporations (their main manufacturing activities are given in parentheses). In 1984, Hyundai and Samsung had sales of $10.3 billion each and were ranked 38th and 39th internationally. Lucky Goldstar, with sales $8.9 billion, was ranked 43rd. Of this, Goldstar's electronics sales were $941 million and Samsung's $835 million in 1983, compared to the total Brazilian informatics output of $1.7 billion by over 100 firms, and compared to India's gross electronics output of $1.36 billion that year, with India's largest firm, Bharat Electronics Limited (BEL), having an output of $147 million.[10] These admittedly crude comparisons give one some idea of the scale of operations of the major Korean firms. By 1990 Samsung, with sales of almost $44 billion, was ranked 18th among the Global (including US) 500; Daewoo, with sales of $25.4 billion, was ranked 43rd, and Goldstar, with sales of almost $5 billion, was ranked 286th. This includes only those Korean corporations whose principal manufacturing activity has been classified as electronics. Among all corporations worldwide whose principal manufacturing is electronics, Samsung's global rank was 5th in 1991, Daewoo's 10th, and Goldstar's 31st.

Samsung, Daewoo, and Goldstar have dominated Korean electronics output and exports. Goldstar, established in 1958, emerged as an exporter of televisions and radios in the 1970s. Samsung Semiconductor and Telecommunications (SST) was established in 1969 and grew to its present size in the export-oriented 1970s as an assembler of consumer electronics, mainly b&w televisions. Since the early 1980s, it has diversified into microcomputers, peripherals, industrial electronics, and ICs. Daewoo had originally been a producer of radios, but it became one of the key players through the acquisition of Taihan in 1981. Hyundai entered electronics only in the early 1980s.

All the four major electronics firms diversified in the 1980s, making determined thrusts into advanced consumer and industrial electronic equipment. They also competed vigorously among themselves in both the domestic and world markets, though mainly on price rather than product innovation—like technological followers—although that is changing. Specialization among firms as promoted by the government in heavy and chemical industries has not been a feature of the electronics industry.

The big four electronics firms have a number of subsidiaries, many of which are joint ventures for the production and export of more advanced products. The growth of the major electronics firms was given a further boost in the mid-1970s by the government's legislative sanction and encouragement of the consolidation of the *chaebol* through the formation of integrated general trading companies (GTCs) "that were given a special status and accorded special privileges. The trading companies were the government's vehicle for decentralizing the administration of export incentives. They were equally its chosen instrument for undertaking the activities needed to strengthen and expand Korea's export marketing capabilities."[11] Korea also had a few large foreign subsidiaries, especially in computers and components, such as IBM Korea, Motorola Korea, and so forth, and a large number of small and medium-sized firms.

The Requirements for International Competitiveness and for an Effective Electronics Policy

The empirical record raises the question of how a country like Korea became a significant electronics exporter even of sophisticated high-technology products, competing against Japan, Europe, and the United States. What roles of the state, the *chaebol*, and the politics of business–government relations have enabled industrial, including electronics, policy to be effective? First, it is necessary to understand the *requirements* for successfully competing in the world electronics market in various phases of development of those technologies and markets.

The policy choices that would have to be appropriate for successfully competing in the world electronics market are the following:

1. *The choice of segments for entry at any given time.* In the Korean case, this would mean components and consumer electronics assembly from the late 1960s and 1970s on, microcomputer assembly in the early 1980s, and semiconductor and telecom exchange design and manufacture in the mid-1980s, building on established capabilities.
2. *The timing of entry into the chosen segments.* This decision is inseparable from the choice of segments for entry. In the Korean case the choice of entry in the early 1970s into components and consumer electronics assembly rather than, say, industrial electronics or component design and fabrication was appropriate in that it anticipated the world boom in these products and coincided with the technological developments in consumer electronics and components that drove MNCs to relocate the assembly stage in low-wage sites.
3. *The correct choice of market agents*—that is, MNCs and joint ventures in components exports and the *chaebol* in consumer electronics and in the joint ventures in components. These choices were appropriate in that they ensured the critical growth requirements, giving access to markets that could provide economies of scale to the industries being set up, access to technology and inputs,

and financial capability to make large, lumpy investments. Organizational economies of scale were also critical.

4. *The phasing of the shift to more advanced products* after the technological, marketing, and financial capability had been developed in simpler products.
5. Related to the third and fourth points above, would be *policies on FDI and imports of technology,* whether with equity participation or via licensing.

Korea seems to have made the right choices in each phase, its policies providing for the technological, financial, and marketing requirements of that phase. We should note that during the early ISI phase the choice of segments for entry were simple consumer electronic items such as radio and TV receivers. These products were helped by the expansion of broadcasting. The choice was appropriate given the timing of entry, since this was prior to the relocation of components assembly by MNCs. These products were easy to master. Korea did not attempt a premature thrust into industrial electronics, nor did it allow foreign firms into this early ISI phase, as it was easy for domestic firms. Exports were not possible at that stage, because domestic assembly of consumer electronics in developed countries was still cost-effective. Nor did Korea set up public enterprises, as there was no question of market failure, their usual rationale. The result was that numerous domestic private firms acquired experience in consumer electronics assembly in an environment of domestic competitive pressures. This laid the foundations for export-led growth.

From the mid-1960s on, it became clear that the domestic market was too small to base further growth on. More importantly, American and other electronic component manufacturers, especially in the semiconductor industry, and a little later consumer electronic equipment vendors had begun to find it cost-effective to get the assembly stage of their manufacturing process done in low-wage locations. Korea's cheap labor, infrastructure, and perceived political stability were important incentives. The Korean government saw this and, as part of general reorientation toward the world market, encouraged MNCs to so relocate. Fairchild made a major investment in 1966, followed by Motorola, Signetics, and other firms. In the 1970s Japanese firms followed suit. Much of this type of investment in components assembly took place in the Free Export Zones.

At the turn of the decade, the Korean government saw that another structural shift was needed. Growth could only be sustained through a shift to advanced consumer and industrial electronics. This, in turn, would require a massive injection of foreign technology and capital. The rules for foreign investment were liberalized but in such a way as to promote joint ventures in which technology transfer to Korean firms would take place. During the 1980s there was again a shift in the role of the state to a more facilitative rather than directive type of intervention, of a more functional than sectoral kind, along with economic liberalization. Yet another phase shift to high-tech R&D-intensive products with appropriate infrastructural support is being attempted in the 1990s.

MAJOR INDUSTRIAL AND ELECTRONICS POLICIES

The Electronics Industry Promotion Law of January 1969 set out explicit policy guidelines to develop an internationally competitive electronics sector mainly by taking advantage of the imperative to relocate labor-intensive component assembly plants for cost-cutting purposes by MNCs. This development was made possible by the Foreign Capital Inducement Law of 1965, amended in 1973, and by the setting up of the Masan Free Export Zone and the Gumi Industrial Complex. The government's fundamental policy concerning foreign capital was "to effectively induce and protect foreign capital conducive to the sound development of the national economy and the improvement of the international balance of payments and to properly manage such foreign capital."[12] Applied to the electronics development plan, this led in the early 1970s to the FTZs and 100% foreign-owned, wholly export-oriented assembly plants, as well as largely export-oriented joint ventures in consumer electronics assembly.

In 1973 three laws were passed which significantly boosted the growth of the electronics industry.

In February 1973, the government announced the Law Regarding Inducement and Management of Public Loans governing long-term foreign loans to the Korean government or private parties. All such loans would require authorization by the Economic Planning Minister and would be available only to certain priority industries, including electronics. The government could lend all or part of the public loans it obtained to institutions dealing in foreign funds or to end-users (here electronics firms). The use of these foreign funds by firms also required authorization; for example, in the case of capital goods imports ministerial authorization would be required and would also constitute an import license. The important point here is that foreign loan capital was funneled to selected export-oriented priority industries and the use of such funds required government authorization.

The second important law was the Revised Foreign Capital Inducement Law of April 1973, amending in some ways the same law of 1965. This law coordinated Korean policy toward all foreign capital inflows, whether loans, direct investment, or technical collaboration. All FDI, lending, or technology inducement had to be authorized by the Minister of the Economic Planning Board (EPB), the apex planning agency. The idea was to "bring the foreign capital inducement system into line with present economic realities and thus make more efficient use of foreign capital for economic development."[13] The key amendment was one that would give priority to joint ventures over wholly foreign-owned export projects. The revised law was aimed at helping Korean firms gain technology, experience, and a share of foreign markets. Another provision of the revised law was the unification of different types of loan contracts, such as cash loans and capital goods loans. Authorization conditions were also made more flexible and discretionary for the Minister of the EPB. All authorizations were to be made by the Minister after consultation with the Foreign Capital Inducement

Deliberation Committee, with the President (Park Chung-hee) having overriding powers. Thus the FDI, loan, and technology import authorization process was unified, centralized, and made discretionary and flexible.

The third and most critical law affecting the electronics industry was the Development Plan for Heavy and Chemical Industries adopted in October 1973. This marked a turning-point in Korean industrialization—a strategic shift toward deepening of the industrial structure. The First Five-Year Plan (1962–66), the first such plan after the advent of the Park regime, concentrated on infrastructural and light industries. The Second Five-Year Plan (1967–71) took the first few steps toward the initiation of heavy and chemical industries.

The Heavy and Chemical Industries Development Plan of 1973 was a concerted push to expand exports of heavy and chemical products (including electronics, especially consumer products). It designated suitable locations for each industry in order to maximize the efficiency of investment through interlocking effects. These industries were designated strategic in that they had important forward and backward linkages, thereby contributing to the development of related industries. Very importantly, they were intended to be export oriented from the very beginning for scale economies. This required a large infusion of foreign loans and/or FDI with technology transfer. The earlier two laws were designed to facilitate this.

The electronics industry was considered labor-intensive but also requiring highly sophisticated technology. Korea was considered to have a comparative advantage in it as electronics exports had risen twentyfold over 1966–72.[14] The Plan adopted measures for the intensive development of electronics dovetailing with the eight-year Electronics Development Plan already in operation.

The Plan showed a preference for joint ventures over wholly foreign-owned firms. It also took an import-substituting step in that it called for the domestic production of materials, parts, and components now being imported, especially for consumer electronics. Another important goal was that of developing consumer electronics as a strategic export industry. In addition, the Plan called for local production of industrial and communications devices that had until then been imported—a bolder and more import-substituting step, but in the form of joint ventures with foreign firms for the introduction of advanced technology.

The Plan's call for relative emphasis on consumer electronics, joint ventures, and the initiation of industrial electronics and components, parts, and materials can be seen as a strategic intervention in the middle of the eight-year electronics development plan to anticipate and develop a comparative advantage in more advanced consumer products for domestic firms dominating consumer electronics exports and thereby to lay the foundation for a later shift toward production and exports of industrial electronics and advanced components by these firms.

The Heavy and Chemical Industries Development Plan established guidelines for obtaining the necessary capital, preferring loans to direct investment, and within the latter category joint ventures, usually on a 50/50 basis, to wholly foreign-owned firms. The desired debt/equity ratio was 70/30, with the EPB able to make discretionary exceptions. This was a very high degree of leveraging, explicable only by the nature of the firms chosen as the market agents for the program—the *chaebol* conglomerates. Only up-to-date technology was to be approved, and all projects were supposed to be internationally competitive, with world-priced products. End-users of planned projects and foreign loans were to be selected in a competitive manner—a policy of picking and supporting winners.

In April 1975, another key law was passed. This was the law establishing GTCs on the lines of the Japanese *sogo shosha* (GTCs).[15] These companies, like their Japanese models, were quite different from firms trading in just a few commodities and from medium and small firms. They were characterized not just by the range of commodities they dealt in, but also by the value and volume of their sales and purchases, their large capital base, their global information and marketing networks, and their ability to extend credit for financing imports and exports. This was prompted by the same economic–environmental factors—that is, the limited size of the domestic market and hence the imperative to export, and the reliance on imported inputs. The GTCs enhanced the competitiveness of Korean industry by being able to make bulk purchases of raw materials and other inputs and thus securing the associated discounts. They were able to overcome the high transaction costs associated with international marketing by smaller firms because they enjoyed organizational economies of scale—that is, scale economies arising from the high fixed costs of global marketing, distribution, and information networks. Both these activities of the GTCs were strengthened by their ability to extend credit for import and export financing—a function of their size and diversified activities. They relieved smaller manufacturing and trading companies of high transaction costs and reduced destructive competition among them.

The Korean government's policy was aimed at promoting the international competitiveness of Korean industry by transforming the existing large-scale trading companies into GTCs and divorcing trading from export manufacturing activities. The criteria announced for firms to qualify as GTCs were minima to be achieved in terms of exports, paid-up capital, number of merchandise items exported, and markets penetrated. For 1975, the company was required to have an export performance of at least $50 million and for 1976 one of at least $100 million. In addition, it was announced that all the firms declared GTCs should go public by 1977 and that the criteria would be made more stringent year by year.

The GTCs were to be given various privileges and incentives. (1) They were to be given priority in participation in international bidding. (2) The government began to issue import licenses exclusively to GTCs for the imports of items restricted to government procurement or to actual users. (3) The GTCs' foreign exchange holdings limit was raised and was to be progressively lifted. (4) They were given preference in membership of export cooperatives. (5) Finally, discriminatory export loan limits for individual firms were introduced, and the GTCs were to be granted exclusive privileges.

The effect of the law creating the GTCs was that it restructured the *chaebol* around the GTCs. The *chaebol* each had one or more trading companies. These companies, at least those belonging to the major *chaebol*, were merged and became GTCs. The other manufacturing and services firms of the *chaebol*, many of whom were big exporters in their respective industries, were grouped around their conglomerate's GTC. Each GTC handled not only the exports and imports of its own *chaebol*'s firms but also those of other unrelated firms, large and small, and that of the government.

The giant electronics firms, those belonging to the Goldstar, Samsung, Daewoo, and Hyundai (since 1984) groups, became major exporters (and importers of inputs) through their *chaebol*'s affiliated GTC. This development helped tremendously to boost the competitiveness of domestic electronics firms due to the advantages of the GTCs in international trade. Significant competitive advantages were derived from this GTC-centered *chaebol*-dominated structure of the electronics industry especially in domestic firm-dominated consumer electronics. In both component assembly joint ventures (where these were not intrafirm operations) and in consumer electronics assembly there are very significant organizational economies of scale. These derive from the reduction of unit costs of inputs due to bulk purchase discounts and the high fixed costs of international marketing. Additionally, there are important externalities, such as better access to world-market and technological information. These benefits were passed on by the GTCs to the electronics giants, being part of the same conglomerate.

The Fourth Five-Year Plan (1977–81) continued development along these lines. However, it became increasingly clear to planners that this strategy's days were numbered. The threat to Korean exports from lower-wage countries such as China, Malaysia, and the Philippines was perceived to be growing. Furthermore, given Korea's dependence on two export markets, the United States and Japan, and on a few labor-intensive products, there was the danger of restrictions on Korean electronics exports by these countries. This acquired greater urgency after the February 1979 quota restrictions on import of Korean color televisions by the United States, French limits on radios, and British limits on b&w televisions.[16] It was also keenly felt that Korean graduation to more advanced, higher-value-added products could be hampered by its crippling dependence on American and Japanese technology and its inability to generate its own. These fears became a reality in 1979–80, when the second oil shock dealt severe blows to the Korean economy, including electronics production and exports. The shock triggered important changes in Korean economic policy, including electronics policy, resulting in a strategic shift in Korean electronics to a third phase of vertical integration backwards into technologically advanced products.

A number of important policy changes to bring about such a strategic shift were initiated. These included the revision of the Electronics Industry Promotion Law in 1981 to promote the production of industrial electronics and more advanced consumer electronics and the development of more advanced technology. Within consumer electronics the emphasis was now to be on more advanced products such as color televisions, videotape recorders, and home computers, helped by the introduction of color broadcasting in 1980 and the reduction of excise duties on domestic sales. The won was devalued by 20% on January 1980 and allowed to float to boost exports.

A thrust into industrial electronics required a number of facilitative measures. The necessary massive infusion of foreign technology would not be forthcoming without equity participation, at least in joint venture form. For even the largest Korean firms to be able to hold their own in such joint ventures, assimilate the technology, and begin to be able to innovate on their own would require further supportive measures—for, after all, competitiveness in industrial electronics depends on the ability to innovate, in turn dependent on the capacity continually to spend substantially on R&D. And the goal of policy was to create an indigenous competitive industry in electronics, freeing Korean industry in the long run from crippling technological dependence.

The Foreign Capital Inducement Act was amended in September 1980, considerably liberalizing the conditions for authorization of FDI.[17] Procedures were simplified (in May 1980), investments up to $10 million were exempted from approval by the Foreign Capital Inducement Deliberation Committee, and repatriation of principal and land acquisition was made easier. From July 1984 on, an automatic approval system for projects that met certain criteria and a negative list system allowing all but listed projects were introduced.

The 1984 amendments also included a comprehensive five-year import liberalization program to expose increasingly competitive Korean firms to international competition and ward off protectionist political pressures in developed countries. This was a policy of moving from ISI to export-promotion in the new advanced electronics segments by phasing out protection as Korean firms became competitive. It was a policy of *industry bias*. Industry bias exists when there are differences in overall effective incentives given to different industries within an overall trade-unbiased regime. This was in line with Korean policy since the 1960s and "discriminated in its treatment between established, internationally competitive industries and new, infant industries that were deemed worthy of promotion."[18] It was analogous to the temporarily protectionist policies of the early 1970s that enabled domestic firms to become competitive in consumer electronics assembly. This was now repeated for color televisions, PCs, VCRs, and so forth, until they became competitive. The instruments used included import quotas, as earlier. The five-year liberalization plan of 1984 also included a graduated liberalization of the capital market, including the phasing out of curbs on foreign banks and securities companies, and portfolio investment in Korean firms, and so forth, and of foreign exchange restrictions (the revised Foreign Exchange Control Regulations of January 1984).

The strategic shift planned in the Fifth Five-Year Plan (1982–86) required a major technological effort by domestic firms with appropriate policy support from the government. Despite the creation of the Ministry of Science and Technology (MOST) in 1967 and several laws promoting scientific and technological development, Korean industry had not begun to invest significantly in R&D until the beginning of the 1980s. Until then, government-run research institutes like KAIST had accounted for the bulk of R&D activity. Nor had foreign firms taken the lead. They did no R&D in Korea and were not even responsible for most of the formal transfer of technology. Technology imports took place mainly through licensing arrangements by local firms.[19] The Fifth Plan set forth the goal of raising the R&D expenditure to GNP ratio from a very low 0.7% in 1980 to 2.0% by 1986. Korea depended heavily on Japanese and American technology, with the United States predominating in overall technology import agreements and royalty payments in the 1980s. There was also considerable technology inflow via joint ventures in ICs and other advanced products. The terms of technology import agreements, including content, duration, payments, and so forth, were authorized by the same agency as foreign capital imports in general and were equally subject to that agency's discretion.

The Korean government's annual budget for science and technology increased at a 15% annual rate over the Fifth Plan. Public sector research institutes accounted for about a third of electronics R&D expenditure in Korea in the early to mid-1980s. Large horizontally integrated private firms also increased R&D expenditures and set up facilities in a major way for the first time in their drive toward vertical integration. Given that the *chaebol* were very large (sales of the order of billions of dollars by the early 1980s), the absolute amounts spent on R&D were adequate to reap the scale economies that exist in R&D—besides which, these expenditures were focused on the new areas such as semiconductor design and fabrication, computers and peripherals, telecom, color televisions and VCRs. The sales volumes of the major firms, their global information networks, and their R&D expenditures gave them the bargaining power to strike better technology transfer deals than would have been the case with smaller, less horizontally integrated firms, as well as the ability to assimilate rapidly the acquired technologies.

The government also facilitated this shift to a third phase with specific industry segment and infrastructure promotion plans such as the Long-term Semiconductor Industry Promotion Plan of 1982–86 and the Computer Industry Promotion Master Plan of 1984. The latter laid out policy for an integrated computer and communications network development plan to promote the development of both industries by the development of a national datacommunication network, the National Administrative Information System (NAIS), to be implemented by the National Computerization Agency (NCA). The Computer Industry Promotion Master Plan was drawn up by the elite science and technology advisory group in the president's office (Blue House) and the Ministry of Communications (MOC). It expanded, in cooperation with the major *chaebol* firms and aided with soft loans, the R&D program of the public sector Electronics and Telecommunications Research Institute (ETRI).

The first two plans mentioned above aimed to promote vertical integration between semiconductors and final products to promote technological spillovers, economies of scale, spreading of risk and synergies, as in the Japanese model. Government control of telecom was used as an instrument to promote semiconductors and, later, computers, through national telecom network expansion, modernization, and technology development. Most chip demand was to be guaranteed to the Korean semiconductor majors, who also entered telecom equipment production in a big way, the government helping renegotiate all existing telecom collaboration agreements with MNCs. However, semiconductors and computers were to be overwhelmingly export oriented for scale economies, mainly on the usual OEM basis, the post-Plaza Accord appreciation of the yen boosting Korean electronics competitiveness, especially in new products such as 256K, 1MB, and 4MB DRAM memory chips, with 16MB and 64MB DRAMs under development. Korea, led by Samsung, introduced 256K DRAM production in 1984, 1MB in 1988, 4MB in late 1989, and 16MB in 1994.

Samsung followed a high-risk leading-edge strategy in competing head-on with the US and Japanese giants in high-volume capital-intensive advanced memory chips after beginning as an OEM and second-source manufacturer dependent on licensed technology. It has achieved recognition of its technological status by its 13-year cross-licensing agreement with IBM in 1989, the first that the latter has struck with a developing-country firm in semiconductors. It gives the two firms free access to each other's entire portfolio of patents in semiconductor design and manufacturing, in return for a one-time fee payment by Samsung. Korea became the world number three semiconductor manufacturer by 1990 and the world DRAM leader in 1994, and Samsung the world DRAM leader in 1992, surpassing Japan and Toshiba, respectively. In computers, the Korean majors began producing PCs, terminals, and monitors in the early 1980s, for OEM exports, overwhelmingly to the United States, building on TV manufacturing skills, aided by compulsory public procurement of Korean machines, domestic content guidelines, and, until 1988, a complete import ban, especially for the greatly expanded national telecom network. They went in for technology alliances with US and Japanese firms and had government R&D backing. Korea also attempted to move beyond OEM PC clones to developing its own 32-bit machines. Until 1986, 8-bit machines dominated exports, but since 1989 32-bit machines have been exported on an OEM basis.

The growth of the national tele- and datacom infrastructure has been vigorously promoted since 1982, for both markets and technology development. Policy has passed through three stages. Up to 1982, the MOC was the government telecom service monopoly, and it also made policy. There was a 5-million-long waiting list for telephones, and service was inefficient. In 1982, telecom service was put under the autonomous Korean Telecommunications Authority (KTA), which set up several subsidiaries for specialized services, including the Data Communications Corporation of Korea (DACOM) in 1982. MOC was limited to policy only. In 1984, the government revised the Telecommunications Law to allow private entry into the database business. KTA and DACOM are the common carriers that own transmission facilities. Massive investments were made in telecom network expansion, increasingly financed by KTA operating profits (from 4.4% in 1982 to 76% by 1987). By 1987, Korea had a fully automated digital-exchange-based nationwide telecom network with direct distance-dialing capability, and 11.8 million lines by 1989, quadrupling since 1980. A major government drive to modernize its own information network led to the growth of the national telecom and datacom networks and value-added services. The 1984 scheme, called Computer-based Integrated National Information System, led to the creation of five national computer-communication networks for government (NAIS), the financial sector, education and research, defense, and health and welfare. From 1988 onwards, there has been a policy shift toward deregulation of private sector entry, especially into enhanced services to promote competition and innovation, and also increasing external liberalization.

Growing rapidly with this increase in domestic demand was the telecom equipment industry. Korea became a significant exporter of terminal equipment, especially telephones. It also developed its own digital electronic switching system (ESS) based on modern Time Division Switching (TDS) technology. The Korean TDX digital exchanges were developed through a policy-led strategic alliance of KTA, ETRI, and four major *chaebol* firms in modular step-up stages from 1981 onwards. The core technology was from Ericsson of Sweden. TDX–1 was developed by 1984, and the TDX–1A, with 10,000 lines, was introduced into the local network in 1986. In 1989, the TDX–10 project for a 100,000 line exchange was launched. Korea became a net exporter of switching systems by 1989.

In the fourth phase of the early 1990s, the infrastructural and R&D support of the state-as-facilitator became even more critical. In 1991, electronics was dubbed a "strategic industry", and the MOC launched an ambitious plan to foster the information and telecom industries focusing on 18 priority technologies in semiconductors, telecom, and computers. In October 1991, a five-year semiconductor equipment plan was launched to produce sophisticated capital equipment in Korea through strategic alliances, one of the aims being independence from US and Japanese suppliers of critical equipment and components, and laying the foundation for future independent brand name exports. In April 1992, the MOC, MOTIE, and MOST jointly unveiled an eight-year national R&D project dubbed G–7, aimed at catching up technologically with the advanced countries; it targeted liquid crystal displays, 64MB and higher DRAMs, and microelectronics generally. A National Strategy for Information and Industry was jointly prepared by these three ministries in 1993, which envisaged massive long-term investment in high-speed broadband datacommunications infrastructure inspired by the US plans for an information superhighway.

The state has deregulated the domestic market and progressively opened up the Korean market to imports, often under pressure from trading partners. Unlike the 1970s preeminence of the EPB, the Blue House, MOC, MOTIE, MOST, and NCA have all played major and coordinated policy roles in the current plans, despite turf battles over telecom equipment and information infrastructure. Technology import has been liberalized, and the former complicated approval system has been replaced by a simple reporting system. FDI has been encouraged as part of high-tech strategic alliance formation and all restrictions on the repatriation of capital and dividends removed, as also on reinvestment in other fields.

The state's role in the 1980s had been of critical importance in R&D support for ETRI to the development of DRAM and ESS technologies. This continues in the 1990s on the post-1984 lines of "big followership," wherein the state will concentrate on supporting collaborative private long-range R&D efforts in critical basic technologies through private–public R&D consortia and infrastructural, informational, and human resource investments, leaving commercialization and marketing to private firms and not intervening in choice of product and other micromanagement decisions. The role of the state will strongly resemble the 1980s MITI model in Japan in the context of the growing globalization of the Korean economy. This transition, ongoing since 1984, has shown results in the success of Korean exports and technology development in progressively more complex products and demonstrates not merely the strategic capacity of the Korean state in Deyo's sense, but the remarkable flexibility, adaptability, and interagency coordination of Korean policy institutions. They were not frozen into a politically insulated and consolidated dominating monolith but began to change well before the democratization of Korean politics that started in 1987.

THE POLITICAL ECONOMY OF INDUSTRIAL AND ELECTRONICS POLICIES IN THE SECOND, THIRD, AND FOURTH PHASES

Korean electronics policy has been extremely successful by most standards in making the transition from ISI to the second phase of labor-intensive assembly industries and then to the third and possible fourth phase of advanced electronics. It has made effective and successful choices with respect to the choice of products, timing of entry, phasing of entry into progressively more complex products, the necessary degree and duration of initial protection, supportive credit policy, public investment policies, regulation of the terms of FDI, loans, and import of technology, and, not least, shifts in the role of the state from "leadership" to "followership" of the market in Wade's terms. What were the political conditions and constraints that made such a policy evolution possible? For this we need to understand the nature of the regime, the decision-making structure of the state, the regime's social base, the relationship of the regime to business groups, and the resulting political interest of the regime in particular economic strategies, or lack of it.

Questions Requiring a Political Explanation

But, first, what are the political problems potentially obstructing the type and timing of the policies that Korea has seen in the past two decades? These are very important questions from a comparative perspective, since many of these problems were dealt with differently by India and Brazil, with very different results for the development of electronics policy and development.

1. How did the regime manage the shift from ISI to export orientation in electronics, when this would mean exposing sheltered domestic firms to international competition?
2. How did the regime manage to use the state-controlled banking system to support the giant firms massively against the much larger number of medium and small firms?
3. Given that the Heavy and Chemical Industry Development Plan of 1973, including electronics, was in part motivated by national security considerations, how did the regime manage to limit entry to those electronics segments where ISI would be efficient and lead to potentially competitive products, without being tempted into industrial electronics and advanced components on grounds of military relevance?[20]
4. Given that foreign equity and technological collaboration authorization were done in a "discretionary command" style, how did the regime withstand the pressures of foreign lobbies for entry into the growing domestic market and greater control?
5. How did the regime guide and control investment decisions by the large firms when the discretionary command style of functioning is obviously prone to corruption? How did it not succumb to either protectionism or the granting of monopolies in various product lines?
6. How did the regime manage the transition to an even more *chaebol*-dominated third phase after the crisis of 1980, with a much greater role for FDI? That is, how did it continue to balance the two with the aim of strengthening the former?
7. How did the regime manage to retreat from the more closely directive intervention of the "big leadership" type to a less sectoral, more functional, "followership" role from the mid-1980s?

The Origin, Evolution, and Character of the Korean Military Regime: A Brief Political History

Before tackling these questions, it is necessary to summarize briefly the political background to Korean development. In May 1961, a military coup was followed by two and a half years of military rule by a junta that initiated the move toward export-led growth. The coup was a response to over a year of social and political unrest following the student uprising of April 1960 and the resignation of the corrupt Syngman Rhee regime. It was a conservative and stability-oriented coup but nationalist in the then Korean context. The military regime faced an economic crisis due to the phasing out of US aid. It moved to reduce dependence on the United States by diversifying diplomatic relations and initiating a policy shift toward labor-intensive exports from 1962 on.

In the political sphere, the regime instituted a "tutelary" democracy in which ultimate power still resided with General Park Chung-Hee, who had taken over in 1961 and remained president of the republic until 1979. However, a national assembly and two parties existed alongside some degree of civil liberties, despite the use of repression to control opposition. This gave way to a period of crisis from 1969 to 1972, marked by worker and student unrest, combined with international crises that strengthened the hand of the already dominant military. The "soft authoritarian" semiparliamentary regime broke down, ending in the emergency of December 1971. This was a period of turbulence marked by unrest in the newly emerged working-class, the student movement, and the political opposition, coinciding with threatening changes in the international environment. Nixon's policy of détente with China destabilized South Korean security by casting doubt on the US defense commitment to the South.

The second international environmental change was in the world economy, where (1) the winding-down of the Vietnam war reduced Korean export earnings, (2) the US imposition of a surcharge on imports and suspension of the dollar's convertibility into gold in 1971 (the Nixon shocks), (3) the US quota on Korean textiles in 1970, and (4) Japanese capital, important in the Zones, wavering in its commitment to Korea, all worsened the domestic crisis of legitimacy and led to the radicalization of sections of the labor and student movements. The regime was faced with a political crisis combined with an economic crisis caused by the exhaustion of, not ISI, but the first phase of labor-intensive export-led industrialization requiring a shift to capital and intermediate goods for exports as well as domestic production.

The Park regime reacted by mobilizing a coalition of military and civilian technocrats and drastically changing the political system to a greatly more repressive, "hard" authoritarianism. The new regime was based on the Yushin (Revitalizing Reform) constitution of 1972, following the emergency decrees of December 1971. The Yushin constitution concentrated power in the hands of the president, who had an unlimited number of six-year terms, drastically reduced the powers of the judiciary and legislature, and gave the president powers to appoint one-third of the members of the latter rubber-stamp body. Repression was dramatically heightened. Nine emergency decrees were issued in the 1970s and martial law declared in 1972 and 1979.

The repressive Yushin regime coincided with the drive to develop heavy industries, from 1973 on. The regime perceived such a strategy as the only growth path feasible in the decade to come. And, unlike in Brazil, the new industries would have to be internationally competitive from the very beginning, due to the smallness of the home market. But for this wages would have to be kept within the bounds of productivity and competitiveness criteria required for exports. This, in turn, would require tight political control of labor. Wage repression of skilled labor in heavy industries was necessary to compensate for the relative technological inefficiency of Korea's new capital goods and chemical industry. This, combined with the fact that the Yushin phase began with greater labor mobilization and unrest, goes a long way toward explaining the "hard" authoritarianism of the 1970s. In October of 1979, President Park was assassinated in a palace coup and replaced by General Chun Doo-Hwan, the regime remaining the same in essentials. A period of prolonged unrest set in again in 1987, culminating in democratic elections in December 1987, in which the government candidate, Roh Tae-Woo, was elected president.

State Structure, State–Society Relations, and Economic Administration: The Political Insulation, Flexibility, and Coordination of Decision-Making

The regime's decision-making on all these policy issues was mainly one of discretionary command over credit, foreign capital, and technology authorization, while tariff and nontariff protection was nondiscretionary in that it applied to all firms. Such discretionary credit and foreign collaboration authorization was used in an "industry-biased" manner both in electronics policy and in general. In electronics, there were industry biases operating at the segmental level in favor of protecting up-and-coming products while de-protecting internationally competitive products.[21] These biases changing in accordance with growing competitiveness and the use of fine-tuned instruments and flexible, responsive decision-making mechanisms, operated throughout the second and third phases.

Korean decision-making mechanisms fell into the category of *discretionary command manipulation*. To clarify what we mean it would be useful, following Jones and Sakong, to classify intervention mechanisms in two ways:[22] (1) in terms of the instruments of intervention (taxes, subsidies, exchange rates, and so forth) and (2) in terms of the pressures brought to bear to secure behavioral compliance. Behavioral compliance can be secured either by *field manipulation* or by *command*. Whereas command implies compulsion, field manipulation may be either *parameter manipulation* or *field augmentation*. The former consists of marginal changes in prices, taxes, interest rates, and so forth, to effect behavior modification by changing the relative payoffs from different courses of action. The latter increases the perceived alternatives. From a market-oriented economic standpoint, field manipulation in either or both of the two ways would seem to be preferred because it employs the market mechanism with its attendant efficiency and conserves monitoring and enforcement costs.

Another dimension of intervention mechanisms, including behavior-compliance mechanisms, is that they can be *discretionary* or *nondiscretionary*. Both field manipulation and command can be either of these. Nondiscretionary intervention applies a general rule or principle to all actors, whereas discretionary intervention discriminates between actors—for example, in the allocation of credit to particular firms in an industry rather than industry-wise fixed terms or fixed terms on a firm size-class basis. Korean intervention has been of the discretionary command type in crucial spheres and in the use of crucial instruments of control, especially credit allocation and foreign capital and technology authorization. It has been nondiscretionary in other spheres, such as exchange rate and tariff policies.

Before one can explain the achievements of Korean policy, one has to understand the nature of the regime and, at another level, its economic policy planning and implementation mechanism. There were four important, observable characteristics of Korean policymaking in the authoritarian period, especially under Park.

(1) The first was that the executive branch—specifically, President Park and his economic secretariat—were overwhelmingly dominant. The legislature was extremely weak and ineffective. All organized interest groups, including, as we shall see, big business, were strictly subordinate to the state. Labor had no power at all. Policy recommendations made at the weekly economic ministers' roundtable were almost automatically ratified at the full cabinet meetings and rammed through the largely rubber-stamp legislature. The EPB Minister had the power to authorize all foreign capital and technology inflows, while credit authorization was done by the government-controlled banks.

(2) The next important characteristic of the policymaking mechanism was the speed and flexibility of policy formulation. This was facilitated by the extreme concentration of decision-making power, unconstrained by formal democratic procedural requirements. However, this has also meant that a large number of hasty, ad hoc decisions were made, only to be modified later. The trade-off seems to have been between carefully deliberated but necessarily slow policymaking on the one hand, and rapid, action-oriented policymaking on the other. While the latter would inevitably mean mistakes, the highly centralized and unconstrained policymaking power structure would allow rapid corrective action. This capacity was well suited to reacting to the exigencies of competing in a rapidly changing world market.

(3) Another important characteristic of policymaking was its "pragmatism" and "openness." In the Korean context, "pragmatism" meant that the goal of export-promotion (say) was sought to be achieved in any way possible using any available instrument. There were no preconceived ideological notions about the desirability or effectiveness of, say, public over private enterprise, small-scale over large firms, state regulation over free competition, the degree of foreign ownership and such like. This resulted in policies which laid down only broad guidelines, easily amended and not rigid rules on, say, the percentage of foreign ownership or of the terms of credit allocation to industrial segments or firms. By "openness" in the Korean context is meant the willingness to listen to a range of opinion (other than labor/left parties) without granting real or even formal decision-making power or even consultative status. This made possible the incorporation of feedback into policymaking, especially for quick corrective action.

(4) Finally, the most important characteristic was "particularism." By this is meant

> the practice of making policy decisions with a low level of generality: for example, with application to only single firm at a particular time. . . . The decision may be codified in a "measure", "directive", "order" or other legal form, but are so often highly specific that the outcome may be characterized as more of a "rule of men" than a "rule of law". The advantage of the system is that it allows fine tuning of decisions in a rapidly changing environment where once-and-for-all general regulations inevitably conflict with a changing environment.[23]

These four characteristics of Korean decision-making taken together explain economic institutional consolidation, interagency coordination, and policy flexibility in electronics. The problem that needs to be politically explained is how and why discretionary command manipulation worked in practice and how it was so effective in surmounting the politically problematic shifts outlined earlier? This requires an analysis of the nature of state power, its social base, its relations to domestic and foreign capital and its level of autonomy from social forces.

Discretionary command is a system of policymaking in which those who hold power—and, in Korea, unaccountable power—are extremely prone to being influenced to use their discretion in favor of particular interests for a consideration: in other words, to rent-seeking behavior. The question then arises: Why did the Korean state not indulge in rent-seeking behavior?

The answer to this question, and to the seven questions about Korean electronics policy posed earlier, lies in the interests of those who control state power and their capacity to translate their power into serving their interests and not those of organized pressure groups. Here we get into the problem of state autonomy from class forces—a part of the general problem of strategic capacity. In the Korean case, this meant from big business—the *chaebol* groups, joint ventures and foreign firms, as well as foreign creditors, labor having been effectively excluded from having a voice. Successful promotion of a competitive electronics industry could have taken place under one or both of two conditions: Either the short-term interests of the above-mentioned forces did not conflict with a phased entry into progressively more advanced electronics products in a world market-oriented way, or the state enjoyed such autonomy that it was able to insulate electronics policy from political pressures.

The relationship between the Park regime and big business was structured in its basic framework after the coup of 1961. Private business had enjoyed a cosy relationship with the corrupt Rhee regime. After Park's coup, the relationship was recast. Most of the top business magnates were arrested in an assertion of absolute state supremacy and autonomy. Eventually, a compromise was worked out that was to define the relationship between the state and the *chaebol* for the entire Park era and beyond, until December 1987.

The crucial characteristic of this relationship was that the state and the *chaebol* were partners in a clearly state-dominated relationship. The reason for the decisive dominance of the state was that the Korean military–absolutist regime did not need the support of business to win elections. Park needed legitimacy and hence stability against potential rivals, for which he needed rapid economic growth. The regime was "dependent upon the success of business but not upon its political support."[24] From this follows the nature of its overall policies toward business and its export-oriented industrial policy, including its policies in the electronics industry.

The Park regime's solid base within the military officer corps gave it the relative autonomy from the *chaebol* required for actions going against their short-term interests or against the particularistic interests of industry groups or individual *chaebol* if necessary. At the same time, the constant pressure on the regime to win and maintain broad popularity, only possible through widening prosperity, impelled it to take a long-term and holistic approach to economic and industrial policy.[25] Thus the regime was faced with the *compulsion* to produce sustained growth, hence the *motivation* to take a long-term view of industrial policy and the *relative autonomy,* and hence the *capacity* to override particularistic and short-term business interests in the process of managing industrialization. Rent-seeking behavior would have retarded growth and damaged popularity and legitimacy. It would perhaps have been rational as immediate maximization of gains if the regime was threatened with collapse or if it lacked the autonomy and instruments to promote growth, which was not the case.

However, on the whole there has been close cooperation between the state and business. Since the regime was dependent on the success of business, it had to consult the latter and frame policies that were conducive to their growth, even while channeling their activities toward regime-determined goals. While the policy framework did not place constraints on the *accumulation* of wealth in private hands, it constrained the *utilization* of wealth. Conspicuous consumption was discouraged, and investment was channeled into state-defined priority areas. Such a framework kept within the basic interests of business even while imposing constraints.

There is no hard evidence from the available literature to build a definitive political economic explanation of the neatly phased electronics policy of Korea. However, it appears possible to construct an explanatory hypothesis for the seven electronics policy issues, requiring a political explanation based on the nature of the relationship between the state and domestic private capital.

(1) As regards the change from ISI to export orientation in the 1960s, it is necessary to point out that protection of uncompetitive industries continued. Foreign capital in the electronics industry entered mainly the Free Export Zones for the assembly of components for re-export. The ISI industry, such as the manufacture of radio receivers and TV sets, received continuing protection, even after the former had become internationally competitive. Trade policy was industry-biased in that it removed protection only for already competitive industries while nursing those not yet so for the future. Lastly, it was precisely during the post-coup transition to export orientation during the 1960s that domestic capital was very weak in relation to the state.[26]

(2) The takeover of the banking system by the state very effectively constrained business and gave the state dominance, by preventing the acquisition of a key form of economic power and independence. This was one of the constraints on the utilization of wealth that the regime imposed. The state enjoyed considerable autonomy in the 1960s as the business–state nexus was broken, and the regime enjoyed solid support in its own constituency, the military. The state's autonomy was further bolstered by the "hard" authoritarianism of the Yushin reforms. Furthermore, the rapid growth of the large firms, exports, and employment allowed the regime to broaden its popularity base. Even small firms prospered in the export-led boom, and thus the potential contradiction between big and small capital present in the regime's use of credit did not come to the fore.

(3) The question of why Korea in its post-1973 phase continued its export-promotion drive in consumer electronics and components assembly without attempting a thrust into more advanced equipment and component fabrication, despite the national security motivation of the Heavy and Chemical Industry Plan, poses a very interesting contrast to India and Brazil in the 1970s. The most likely explanation seems to be that the Koreans found it most immediately and pressingly necessary to create the wherewithal for heavy arms production, the staple of warfare. Electronic equipment was a secondary consideration, and the economic—that is, balance of payments—motivation of the 1973 shift dominated the electronics component of the Plan. In addition, there was probably a feel for reality in that it was realized that creating an advanced electronic equipment industry was not possible for Korea in the early 1970s. Furthermore, even a US withdrawal would probably not have resulted in loss of access to US arms on a purchase, if not aid, basis. Thus Korea was able to avoid becoming bogged down in inefficiency by prematurely attempting an advanced equipment industry.

(4) The entry of foreign capital in electronics production aimed at the domestic market in products that compete with domestic firms clashes with the latter's interests. Here state interests in promoting the development of indigenous capabilities and the interests of domestic firms simply coincided, at least in the 1970s, as long as the Korean domestic market was not large enough to attract foreign capital seriously, and Korean firms exported consumer electronics on an OEM basis for foreign producers and mass merchandizers. In the 1980s, Korean firms made a bid to compete in more advanced products requiring massive capital and technology imports in the form of joint ventures, and the Korean market grew sizeable enough to attract foreign capital.

Explaining the state's ability to promote domestic and ward off foreign capital now becomes trickier. Discretionary command is obviously prone to corruption and political influence. While foreign firms made major inroads into Korean industry in the form of joint ventures in the 1980s, these were either overwhelmingly export-oriented, as in computers, or designed to buttress domestic capability in semiconductors and other high-tech areas. The OEM strategy was followed in computers, terminals, and monitors, and joint ventures facilitated it. Therefore, although there was greater foreign equity participation, Korean firms used this route to the world market in progressively more advanced products—for the first time even under their own brand names. It is not clear whether this was a situation of greater foreign capital dominance. There is no clear explanation, but it appears that domestic firms had access to the organ of discretionary authorization of foreign capital and technology import and influenced it to bolster their new export thrust.

(5) The state did use discretionary credit allocation and other instruments to protect domestic firms and in some cases—for example, Samsung Semiconductor in 256K DRAMs—grant de facto monopolies over certain products.[27] However, the *chaebols*' influence over the discretionary command mechanism did not lead to inefficiency behind protectionist barriers or monopoly because almost all electronics products were overwhelmingly dependent on the world market. Therefore, the *chaebol* had a stake in competitiveness. Protecting or granting monopoly in the domestic market would shield only a small part of revenues in any product line unlike in the case of a predominantly ISI regime, as in India or even in Brazil.

In some cases of de facto monopoly, such as Samsung in 256K DRAMs in 1984, the capital intensity and hence volumes required made a single producer the only viable national strategy. But this "national champion" was forced to face global competition due to the fact that the markets for memory chips lay abroad. Thus in the electronics industry discretionary command did not lead to ISI-type inefficient industrialization. However, one is forced to admit that this is a tentative explanation. It is not wholly satisfactory, for, after all, in India and Brazil protected firms produce for the domestic market on a comically small scale. The explanation has to include the extreme political insulation of the discretionary command authority from the *chaebols* and other pressure groups, so as to be able to plan rationally, going by threshold size and scale economy criteria.

(6) We have partially discussed the possible explanation for the state's achievement of this transition in the fourth point above. But as regards the still greater role of the *chaebol* in this third phase, there is probably a relatively straightforward economic explanation. Entry into the more capital-intensive segments could be effected only by the largest firms. Hence, once competitive pressures forced the *chaebols* to make moves in this direction and the state decided to support such a shift for reasons of anticipated comparative advantage and anticipated loss of the same in labor-intensive products, the *chaebols* would be the natural candidates for state support. The result would be still greater concentration in the electronics industry, but smaller firms would be in no position to contest this.

(7) The shift from directive to facilitative intervention reveals the state's *policy flexibility*. This can be explained by the fact that (a) the state (and its banks) never allowed itself to be captured by the *chaebols*; it was the dominant partner throughout, and its directed credit, protection, and subsidies were always export-conditional. Therefore, it could withdraw or shift the type of support being provided without protection- and subsidy-dependent firms being able to exert

effective counterpressure. The four characteristics of policymaking discussed earlier also contributed to policy flexibility. (b) As Nayar has argued, the inverse relationship between the degree or depth of state intervention and policy flexibility probably explains a great deal. No major public sector industry of the Indian or Brazilian kind was created in electronics. Therefore, it did not face the problem of withdrawing from direct production employing a large labor force. The historical path of intervention followed did not create interests that could dominate the state and constrain policy changes. Ideologically, too, there was no path-created constraint on policy shifts because the dominant ideology was utterly result-oriented.

To sum up, we have tried to construct political explanations based on the nature of the state and its relationship with (autonomy from) the principal organized forces—that is, those of big business—to explain certain difficult issues in how and why over time electronics policy took the shape that it did in the face of what appear to be *political* obstacles. Our explanations, it must be emphasized, are tentative and conjectural. What is clear is that the state had all the ingredients of strategic capacity in the extended sense—autonomy, economic institutional consolidation, and inter-agency coordination and the policy flexibility to effect shifts in intervention patterns in accordance with the requirements of competitiveness. The state elite also had the political motivation to define international competitiveness as its strategic priority and hence make it central to its politico—economic grand strategy.

NOTES

1. This section draws heavily on Martin Bloom, *Technological Change in the Korean Electronics Industry* (Paris: OECD, 1992); Ashoka Mody, "Institutions and Dynamic Comparative Advantage: The Electronics Industry in South Korea and Taiwan," *Cambridge Journal of Economics*, 14 (1990): 291–314; Robert Wade, *Governing the Market*; World Bank, *Korea: Managing the Industrial Transition* (Washington, DC: World Bank, 1987) Vol. 2, pp. 189–225; Peter B. Evans and Paulo Bastos Tigre, "Going Beyond Clones in Brazil and Korea: A Comparative Analysis of NIC Strategies in the Computer Industry," *World Development*, 17, no. 11 (1989): 1751–1768, and "Paths to Participation in 'Hi-Tech' Industry: A Comparative Analysis of Computers in Korea and Brazil," *Asian Perspective*, 13, no. 1 (1989); Byung Moon Byun, "Growth and Recent Development of the Korean Semiconductor Industry," *Asian Survey*, 34, no. 8 (1994): 706–720; Charles Edquist and Staffan Jacobsson, "The Integrated Circuit Industries of India and the Republic of Korea in an International Techno-Economic Context," *Industry and Development*, 21 (1987): 1–62; Chang-Ho Yoon, "International Competition and Market Penetration: A Model of the Growth Strategy of the Korean Semiconductor Industry," in Gerald K. Helleiner (ed.), *Trade Policy, Industrialization and Development* (Oxford: Clarendon Press, 1992); Cae-One Kim, Young Kon Kim, and Chang-Bun Yoon, "Korean Telecommunications Development: Achievements and Cautionary Lessons," *World Development*, 20, no. 12 (1829–1841); Korea Exchange Bank, *Monthly*

Review, various issues; *Business Korea*, various issues; *Electronics Korea*, various issues; Electronic Industries Association of Korea (EIAK), *Statistics of Electronic and Electrical Industries: Production, Export, Import* (Seoul: EIAK), various years; Elsevier Advanced Technology, Oxford, UK, Benn Electronic Publications, *Yearbook of World Electronics Data*, various years.

2. Mody, "Institutions," 3–4 and Table 1.17; Jee-man Chung, "The Electronics Industry in Korea," 19, Table 2.

3. World Bank Country Report, "Korea: Development in a Global Context," Annex 3, Table 5.2, p. 208.

4. Ashok Parthasarathi, "Electronics in Developing Countries: Issues in Transfer and Development of Technology," *UNCTAD* (1978): 27; Charles Edquist and Staffan Jacobsson, "The Integrated Circuit Industries of India and the Republic of Korea in an International Techno–Economic Context," *Industry and Development*, 21 (1987): 1–62, especially 23.

5. Table 5.1; EIAK data; Salamao Wajnberg, "The Brazilian Microelectronics Industry and Its Relationship with the Communications Industry," UNIDO/IS.546 (8/5/1985): 121–124.

6. Elsevier Advanced Technology, Oxford, UK, sources of data in Tables 5.3 to 5.6 and publishers of Benn Electronic Publications, *Yearbook of World Electronics Data*, various years, gives comparability-adjusted figures for all countries. It must be noted that EIAK figures for Korean production, market size, exports and imports—the source for all figures on Korean electronics in this section—are higher than Elsevier figures by a relatively small margin because they include household electrical appliances and other products not included in the latter source's comparability-adjusted definitions. However, the rough order of magnitude and the trends revealed are not contradictory.

7. *United Nations, Yearbook of Industrial Statistics* (New York: United Nations), various years.

8. Ibid.

9. Frischtak, Claudio (1986), "The Information Sector in Brazil: Policies, Institutions and the Performance of the Computer Industry" (Washington, DC: World Bank, 1986), p. 10.

10. Bharat Electronics Limited, *Annual Report* (1983–84).

11. Howard Pack and Larry E. Westphal "Industrial Strategy and Technological Change," *Journal of Development Economics,* 22 (1986), p. 96.

12. Korea Exchange Bank, *Monthly Review*, 7, no. 4 (April 1973): 1–7.

13. Ibid., p. 1.

14. Ibid., p. 8.

15. Korea Exchange Bank, *Monthly Review*, 9, no. 11 (November 1975): 1–7.

16. Korea Exchange Bank, *Monthly Review*, 13, no. 11 (Nov. 1979): 12.

17. Sang-Yil Chun, "Direct Foreign Investment in Korea," Korea Exchange Bank, *Monthly Review*, 14, no. 11 (November 1980): 1–18.

18. Pack and Westphal, "Industrial Strategy," p. 94.

19. For the information in this paragraph, see Jung-Hyun Kim, "Recent Developments in R&D in Korea," Korea Exchange Bank, *Monthly Review*, 19, no. 12 (December 1982): 1–15.

20. For full development of the argument that the 1973 shift was motivated by national security considerations, see In-won Choue, "The Politics of Industrial Re-

structuring: South Korea's Turn Toward Export-led Heavy and Chemical Industrialization 1961–74," Ph.D. dissertation, Department of Political Science, University of Pennsylvania, 1988.

21. Pack and Westphal, "Industrial Strategy," pp. 94–96.

22. The following discussion draws heavily on Jones and Sakong, *Government, Business.*

23. Jones and Sakong, *Government, Business*, pp. 64–65.

24. Ibid., p. 68.

25. For a full development of this argument, especially for the 1973 shift to heavy industry, but also for the export politics model, see In-won Choue, "The Politics of Industrial Restructuring."

26. For analyses of the post-coup transition, see Jones and Sakong, *Government, Business*, pp. 280–282; David C. Cole and Princeton N. Lyman, *Korean Development: The Interplay of Politics and Economics* (Cambridge, MA: Harvard University Press, 1971), pp. 34–50.

27. Edquist and Jacobsson, "The Integrated Circuit Industries," pp. 53–55.

4

The Political Economy of Import-Substitution in the Brazilian Electronics Industry

The electronics industry in Brazil, unlike in Korea and India, was dominated from the inception by foreign firms. In the past two decades, the policy intervention of the Brazilian state in the development of the industry has been the story of the "Brazilianization" of the ownership, control, and technological capability of the industry. This has been the case, most pronouncedly, in the informatics (computers, software, telematics) and telecommunications (manufacturing and services) segments. State promotion of the industry has been predominantly ISI-oriented. The general thrust of policy has been to promote Brazilian-controlled firms and progressive Brazilian control of established foreign firms, with a view to indigenization of technological capacity.

Brazil's electronics development and its strategy thus make for an interesting contrast with both Korea and India. Unlike Korea, its main thrust was ISI, and it started with a large foreign presence dominating both the domestic market and exports. It resembled India in that it made a determined thrust toward indigenization of telecom and computers from the mid-1970s rather than a Korean-style phased approach. However, unlike India, it did not try to restrict foreign capital and technology drastically, nor give state-owned enterprises the leading role in manufacturing. Rather, it was a strategy aimed at developing an indigenously owned and controlled informatics and telecom industry within an ISI framework without ousting foreign firms but gradually bringing them under national control, molding their behavior, and acquiring their technology.

The questions that arise are: Why did Brazil choose an ISI and industrial electronics-oriented strategy? Why were foreign firms allowed a considerable presence even during an overall nationalist shift? Why was the public sector not given a leading role at a time when state enterprise was spearheading Brazilian ISI in heavy industry? What were the forces and constraints behind such policy choices? The answers to these and other questions of political economy lie in the nature of political power in Brazil during the military regimes following the 1964 coup and the new democratic regimes since 1985.

BRAZILIAN ELECTRONICS PRODUCTION: THE EMPIRICAL RECORD

Overview of the Industry

The Brazilian electronics industry can be classified into consumer electronics, informatics (computer hardware, software and services), telecommunications (telecom equipment manufacture and services), components, and aerospace/defense. The first is treated differently from the rest, administratively and policy-wise. On the last we have no separate reliable quantitative information, except for the figures under the entry, "Communications and Military equipment" in the *Yearbook of World Electronics Data.*[1] It is integrated with the aerospace/defense industrial complex. The core of the industry consists of the informatics and telecom segments, these having been the focus of promotional policy interventions.

Brazil's electronics industry has been substantially larger than India's throughout. Table 4.1 gives segment-wise production figures and Tables 5.3 to 5.6 contain comparisons with Korea and India. Gross electronics output was about $325 million in 1972 with a 60–40 divide between consumer and industrial electronics.[2] It rose to $600 million in 1976 and to $4,148 million in 1981. By 1987, it had fallen behind Korea, its electronics output being $11.33 billion, compared to Korea's $13.68 billion, using comparable figures. By 1989, its domestic market was smaller than Korea's—$13.67 billion, compared to Korea's $13.76 billion. By 1993, its output was $12.79 billion, compared to Korea's $29.16 billion, and its domestic market was $14.5 billion, compared to Korea's $17.9 billion. However, consumer electronics had shrunk to 16–18% of output for the years 1985–93 from 40% in 1972, and its share of market size was 14–16% for the same years. By contrast, the share of EDP in output rose from 33% to 39% over 1985–90, declining slightly by 1993; its share of market size hovered around 33–36% over 1985–93. Telecom share of both output and market size hovered between 9% and 12% over 1985–93, taking industrial electronics equipment (including office equipment, control instrumentation, medical, communication, and military electronic equipment) from 60% to 65% of output and domestic market size over 1985–93. Exports have been only a small proportion of production—6–8% over 1985–93—and have been mainly from MNC-dominated consumer and component assembly segments.

We summarize briefly the main trends in the principal segments in Brazilian electronics—computers (called informatics) and telecom.

Informatics was dominated by foreign firms, principally IBM, until the mid-1970s. Tables 4.2 to 4.4 give aggregate and segment-wise shares of sales between Brazilian national firms and other (foreign) firms in electronics. Since then, as a result of a state-planned policy initiative to develop a nationally controlled EDP industry, the segment has taken off and experienced runaway growth. The number of installed computers of all size classes increased from a mere 506 in 1971 to 6,060 by 1977, to 14,069 by 1981, and to 810,000 by 1986.[3] Of the last figure, some 98% are estimated to have been locally manufactured. The explosive growth after 1980 is due to the microcomputer boom. Earlier, during the international era of the mini between 1973 and 1981, this category of system showed the highest growth rate per year (62%) of all six size classes.[4]

In telecommunications, Brazil made striking progress in indigenization of production and technology as a consequence of policy initiatives beginning in 1974. Before that, the telecom infrastructure was unplanned and inefficient. In the mid-1960s, there were 1.3 million telephones. National telex coverage was very limited, with only 1,000 terminals in the country. The number of telephones grew from 1.66 million in 1968 to 7.8 million direct exchange lines by early 1988, and to around 10 million by the end of the decade.[5] Correspondingly, telephone coverage per 1,000 population increased from 19 to 60 over the same period.[6] This extremely rapid growth was largely due to the nationalization of the telecom network under the public enterprise TELEBRAS in 1972. There has been impressive progress in updating and expanding the telecom transmission network, including the provision of new tele- and datacommunications services, led from 1980 on by EMBRATEL, TELEBRAS' long-distance subsidiary. Transmission was predominantly by microwave line-of-sight relays yielding to satellite transmission and fiber optics.

As regards equipment manufacturing, until the early 1970s four MNCs—Ericsson, Standard Electric SA (an ITT subsidiary), Siemens, and Nippon Electric Co. (NEC)—were dominant, with over 90% of total sales.[7] From 1974 onwards, the newly formed Ministry of Communications (Minicom) and its enterprise TELEBRAS carried out a series of nationalist policies that gradually increased Brazilian control and molded MNC behavior to conform to national objectives. This included progressive local manufacture of electronic—including digital—exchanges and local R&D. The result was a rapid increase in Brazilian production of telecom equipment and a corresponding fall in imports. The import substitution ratio for telecom equipment increased from 80% in 1975–76 to approximately 90% in 1983, considerably more than for other electronics segments at the time—and this despite the introduction of the new electronic switching technology in 1980 and its becoming the bulk of the switching market by 1984. This must certainly be counted as a signal achievement of local technological effort.

Table 4.1: Gross Sales of Brazilian Informatics Sector (US$ million)

Segment	*1980*	*1981*	*1982*	*1983*	*1984*	*1985*	*1986*	*1987*	*1988*	*1989*	*1990*
EDP	860	1,040	1,508	1,487	1,728	2,115	2,126	2,578	3,373	4,337	3,719
Teleinfo						437	636	617	945	1394	1,478
IndAuto					86	101	199	294	281	464	341
Microel.							209	242	327	469	310
Software							167	208	240	389	351
DigInst.					19	25	55	77	88	110	135
Total	860	1,040	1,508	1,487	1,833	2,678	3,392	4,016	5,254	7,163	6,334

Source: Brazil: Department of Informatics and Automation Policy (DEPIN), *Panorama do Setor de Informatics*, Series Estatistics, Vol 1, No 1 (September 1991), p. 7.

Note: EDP = Electronic Data Processing (computers); Teleinfo = telecom; IndAuto = Industrial Automation; Microel. = microelectronics; DigInst. = digital instrumentation. The broadly defined informatics sector above, includes telecom equipment, control instrumentation, medical and industrial electronics (under Industrial Automation and Digital Instruments), and microelectronics, including semiconductors. However, it does not include sales of technical services by informatics companies, which is a significant minority of their sales (about a quarter) since 1985. when it was disaggregated. Before 1984, informatics figures included only EDP activities. These figures do not coincide with but do not contradict the general picture of Brazilian electronics revealed by Elsevier Advanced Technology, Oxford, UK (figures in later tables).

Table 4.2: Gross Sales of Companies in Brazilian Informatics Sector (in US$ million and %)

Types of Company	*1980*	*1981*	*1982*	*1983*	*1984*	*1985*	*1986*	*1987*	*1988*	*1989*	*1990*
Brazilian National Companies	280	370	558	687	952	1,400	2,081	2,378	2,811	4,243	3,822
(%)	(32.6)	(35.6)	(37.0)	(46.2)	(51.9)	(52.3)	(61.4)	(59.2)	(53.5)	(59.2)	(60.3)
Other companies	580	670	950	800	881	1,278	1,311	1,638	2,443	2,920	2,512
(%)	(67.4)	(64.4)	(63.0)	(53.8)	(48.1)	(47.7)	(38.6)	(40.8)	(46.5)	(40.8)	(39.7)
Subtotal	860	1,040	1,508	1,487	1,833	2,678	3,392	4,016	5,254	7,163	6,334
Grand Total	860	1,040	1,508	1,487	1,833	3,592	4,578	5,362	6,897	9,843	8,621

Source: Brazil: Department of Informatics and Automation Policy (DEPIN), *Panorama do Setor de Informatics*, Series Estatistics, Vol 1, No 1, September 1991, p. 16.

Note: Other companies = foreign-controlled companies. The broadly defined informatics sector includes telecom equipment, control instrumentation, medical and industrial electronics, and microelectronics, including semiconductors. However, it does not include sales of technical services by informatics companies, which is a significant minority of their sales (about a quarter) since 1985, when it was disaggregated. Before 1984, informatics figures included only EDP activities.

Table 4.3: Gross Sales of Brazilian National Companies in Informatics Segments

Segment	*1980*	*1981*	*1982*	*1983*	*1984*	*1985*	*1986*	*1987*	*1988*	*1989*	*1990*
EDP	280	370	558	687	847	1,082	1,242	1,375	1,545	2,312	1,920
Teleinfo						192	359	378	592	867	1,080
IndAuto					86	101	199	294	279	462	340
Microel.							109	132	195	289	195
Software							117	122	112	203	152
DigInst					19	25	55	77	88	110	135
% of industry	(32.6)	(35.6)	(37.0)	(46.2)	(51.9)	(52.3)	(61.4)	(59.2)	(53.5)	(59.2)	(60.3)
Total	280	370	558	687	952	1,400	2,081	2,378	2,811	4,243	3,822

Source: Brazil: Department of Informatics and Automation Policy (DEPIN), *Panorama do Setor de Informatics*, Series Estatistics, Vol 1, No 1 (September 1991), p. 18.

Note: EDP = Electronic Data Processing (computers); Teleinfo = telecom; IndAuto = Industrial Automation; Microel. = microelectronics; DigInst = digital instrumentation; % of industry = Brazilian national companies' share of total sales in the industry (excluding sales of technical services).

Table 4.4: Gross Sales of Other Brazilian Companies in Informatics Segments

Segment	*1980*	*1981*	*1982*	*1983*	*1984*	*1985*	*1986*	*1987*	*1988*	*1989*	*1990*
EDP	580	670	950	800	881	1,033	884	1,203	1,828	2,025	1,799
Teleinfo						245	277	239	353	527	398
IndAuto							0	0	2	2	1
Microel							100	110	132	180	115
Software							50	86	128	186	199
DigInst	—	—	—	—	—	—	—	—	—	—	—
% of industry	(67.4)	(64.4)	(63.0)	(53.8)	(48.1)	(47.7)	(38.6)	(40.8)	(46.5)	(40.8)	(39.7)
Total	580	670	950	800	881	1,278	1,311	1,638	2,443	2,920	2,512

Source: Brazil: Department of Informatics and Automation Policy (DEPIN), *Panorama do Setor de Informatics*, Series Estatistics, Vol 1, No 1 (September 1991), p. 18.

Note: EDP = Electronic Data Processing (computers); Teleinfo = telecom; IndAuto = Industrial Automation; Microel. = microelectronics; DigInst = digital instrumentation; other Brazilian companies = foreign-controlled companies; % of industry = other Brazilian companies'share of total sales in the industry (excluding sales of technical services).

The Brazilian consumer electronics industry consisted predominantly of TV and radio receivers (including auto radios and radio-cassette recorders). In the early 1970s it was larger than Korea's and India's. However, due to exports, Korean output rapidly overtook Brazil's in value and volume (Table 4.5 for output volumes of major products in comparison with Korea). By 1976, Korean production of TV sets exceeded Brazil's. However, its domestic consumer electronics market has always been much larger than India's, and even in 1990 it was still larger than Korea's due to its much larger population (about 150 million in 1990, compared to Korea's 43 million) despite lower per capita income. Brazilian and Korean production of radios, TV sets, and telephones over 1973–91 is shown in Table 4.5. Brazilian production of simple radio receivers peaked at 1.3 million in 1973 and then declined alongside rising production of

Table 4.5. Brazilian and Korean Production of Radios, Televisions, and Telephones, 1973–1991 (1,000 units)

	Brazil			*Korea*		
Year	*Radio*	*Television*	*Telephone*	*Radio*	*Television*	*Telephone*
1973		1,480		3,272	816	
1973		1,991		3,692	1,164	
1975		1,607		4,464	1,225	
1976		1,916		6,717	2,290	
1977		2,078		6,404	2,990	
1978		2,422		4,768	4,826	
1979		2,747		4,772	5,867	
1980		3,554		3,972	6,819	
1981		2,517		5,086	7,524	
1982		2,354		5,925	6,113	2,294
1983	5,010	1,857	1,134	6,719	7,641	6,936
1984	5,722	1,744	1,361	7,709	9,730	4,358
1985	6,418	2,187	1,388	1,031	7,849	5,907
1986	8,519	3,034	1,353	1,493	11,799	8,105
1987	8,676	2,902	1,801	1,425	14,922	9,523
1988	7,632	2,722	1,247	1,414	14,820	10,293
1989	7,210	2,920	816	1,112	15,469	10,875
1990	5,151	3,196	1,341	1,462	16,184	10,373
1991	5,317	3,265	1,005	830	15,986	9,536

Source: United Nations, *Yearbook of Industrial Statistics,* 1991, Vol. 2 (New York: United Nations).

Note: "Radios" includes radio-cassette recorders, car radios, and other types, including composite systems.

more sophisticated consumer products such as tape decks, calculators, and stereo equipment.

The total Brazilian consumer electronics market hovered in the range of $1.99 to $2.29 billion during the stagnant 1980s. Of this, the radio and TV receiver market was in the range of $1.5 billion to $1.81 billion. Color television had a rising share since the introduction of color broadcasting in the early 1970s, and it dominated consumer electronics in the 1980s, while b&w television declined. Brazilian production was aimed predominantly at the domestic market. Production, especially of color televisions, was mostly in the Manaus Free Export Zone. Production was initially mainly assembly of imported components, but component production increased in Brazil over the 1980s. However all ten color TV producers remained well below plant and firm-level minimum efficient scale at the end of the 1980s.[8]

The Brazilian electronic components industry also grew rapidly, as would be expected from the growth of equipment production. However, it has not kept pace with the latter, which depended heavily on imported components, falling from over a quarter of electronics output to just over a sixth over 1985–90, despite progressive ISI in less advanced components, especially printed-circuit boards, TV picture tubes, and less complex semiconductors. However, Brazil has developed considerable technological capability, especially in telecom ICs and in ASICs in its two principal research centers for semiconductors, TELEBRAS's CPqD and the Centre for Informatics Technology (CTI).[9]

From their inception, MNCs have had a dominant presence in all segments of electronics. However, their share has been gradually reduced in informatics and telecom due to initiatives since the mid- to late 1970s. Over 1980–90, the share of national firms rose from 40% to 80%. By the mid-1980s, small computers were 100% produced by Brazilian-owned firms, although MNCs continued to dominate mainframes. In telecom, the share of MNCs was forced down since the mid-1970s, and they were forced to upgrade the technological level of their locally made products. In components and consumer electronics, MNCs remained dominant. At the end of the 1980s, all but one of the ten color TV producers were joint ventures or MNC subsidiaries dependent on technology and component supply. From the early 1980s on, major Brazilian informatics producers began to enter the semiconductor and other component fields. However, MNCs often exercised technological control, even without controlling ownership, as in 32-bit superminicomputers, digital switching equipment, and semiconductors.

This outline of the MNCs versus national firms in Brazilian electronics needs to be complemented with an account of Brazil's technological progress in electronics. There are two aspects to this: (1) the technological progress achieved by national firms and laboratories; (2) the transfer to Brazil by MNCs of the latest technology under policy pressure from the state, since this contributed critically to the Brazil's technological base.

Electronics Technology and Brazil's Technology Policies

Technology development was a prominent feature of the development of Brazil's electronics industry. But first for an historical overview of technology policies in Brazil:[10] In 1951, the CNPq (National Research Council) and the CAPES (Campaign for Improvement of High Manpower) were founded, and in 1964 a Special Fund for Technology (FUNTEC) was started within the BNDE. When science and technology become an explicit policy objective after 1968, the SNDCT (National System of Scientific and Technological Development), a set of plans, and FNDCT (National Fund for Scientific and Technological Development), to finance the former, were established. The CNPq, the premier organ of the SNDCT from 1974 on, was brought under the Planning Secretariat (SEPLAN) and given executive powers.

Import of technology was regulated as part of the law regulating import of foreign capital and started as early as 1962. In 1971, INPI (National Institute of Industrial Property) was set up and took over the role of screening technology transfer agreements. It played a nationalist role in strengthening licensees' bargaining power, resisting restrictive clauses, and so forth.

In electronics, the premier R&D institution is TELEBRAS's center, the CPqD (Center for Research and Development). Since its founding in 1976, this institution has done pioneering work in Brazil in the fields of telecom equipment (switching, transmission, and peripheral) and semiconductor components geared to the needs of the telecom equipment program. CPqD's most impressive achievement has been the development of a whole family of exchanges. These include four electromechanical exchanges, one electronic (SDS) exchange, and a sixth, the TROPICO exchanges, a family of TDS technology-based exchanges. From 1985 on, TELEBRAS procurement for smaller exchanges up to 4000 lines was reserved for CPqD's indigenously developed TROPICO R small exchanges produced by local firms. CPqD also has active R&D programs in digital transmission, fiber optics, peripheral equipment, satellite communications, and components. These programs collaborate actively with various state and private enterprises and university labs, including the University of Campinas (Unicamp), CETUC (Center for Telecom Studies) of the Catholic University of Rio, the LME (Laboratory for Microelectronics) of the University of São Paulo, and the LED (Laboratory for Electronic Devices) of the University of São Paulo. MNCs, in a turnaround, also joined these R&D efforts.

In informatics, Brazilian firms have made important technological progress in the microcomputer and software areas under the market reserve policy. COBRA, the state enterprise, set the pace in the 1970s with the development of mini systems. Local product innovations at the systems architecture level, using standard off-the-shelf components, are common. Locally developed software is also doing well, including finding a market abroad. However, at the level of superminis and larger systems Brazil continues to be heavily dependent on licensed technology.

In components, technological dependency is acute, but in recent years there has been an indigenous R&D effort led by state institutions and national firms. CPqD is at the forefront of the national R&D effort in components too, especially for telecom, as is the CTI since 1982. Other leading R&D centers are located at the University of Campinas, the University of São Paulo, and the Federal University of Rio de Janeiro. The general thrust has been toward developing LSI/VLSI design—and later full-process—capability for semi-custom and custom chips, as these do not require the scales that standard memories and logic chips do and could be competitively designed and produced in Brazil. What is significant is that all R&D efforts have come from state enterprises, institutions, and national firms, and only later from MNCs under policy pressure.

MAIN FEATURES OF BRAZILIAN ELECTRONICS DEVELOPMENT FROM THE EMPIRICAL RECORD

Several features of the Brazilian electronics industry's development are noteworthy:

(1) The growth of the industry, over the whole period, overall and in all its segments, has been slower than that of Korea but faster than that of India, except for the second half of the 1980s. However, the difference between the growth rates of the domestic market size of Korea and Brazil is much narrower than that of output. Korea's domestic market remained smaller than that of Brazil until 1990. In EDP, and in industrial electronics in aggregate, Brazil's domestic market remained twice as large, and 30% larger than that of Korea in 1993. Korea's output lead and Brazil's falling behind was due to the facts, respectively, that the Brazilian industry relies almost wholly on the domestic market, whereas Korean output shot ahead in the 1980s due to a successful penetration of the world market, especially for components and consumer electronics.

(2) Brazil's development has been of an ISI type. Exports have been very small—only $725 million, compared to Korea's $15,682 million in 1990, and 1083 million compared to 19,646 million in 1993. Even this has been mainly of consumer electronics and components assembly products, largely by MNCs and joint ventures, unlike Korea's domestic firm-generated exports.

(3) The industrial electronics segments (informatics, telecom, and other equipment) have been the leader in Brazil's development. The ISI thrust has been in this area, and its weight in Brazil's output had risen from an already high 40% in 1972 to 65% by 1990. This shows that industrial equipment grew much faster than did consumer electronics, and that therefore there was no phasing-in of complex production, as there was in Korea. Brazil's industrial electronics-led ISI resembled the Indian model. Brazilian ISI, in all segments, suffered from a lack of scale economies and overdiversified firms, with all their techno-economic consequences for competitiveness, as in India, but to a lesser degree due to a larger home market.[11]

(4) Brazilian ISI was carried out mainly by domestic private firms, but also to a significant degree by local MNC subsidiaries. The state's role in direct production remained very small. Only COBRA in informatics (specializing in minis in the 1970s) represented the state in a significant way, and it, too, was a private–state joint venture and lost its top position among national firms in the 1980s. The largest firm in computers remained IBM do Brasil. MNCs dominated even TV production. In telecom, the leading firm producing switching systems remained foreign-owned but nationally controlled through voting shares control.

In its choice of market agents, the Brazilian model differed significantly from the Korean on the one hand and the Indian on the other—from the former in the heavy presence of MNC subsidiaries in home-market-oriented production, and from the latter in the near-absence of the public sector in production (as against R&D or services). However, the private electronics firms resemble those of Korea more than those of India, in that they tended to be part of industrial/banking conglomerates from their inception rather than independents—a pattern that began to emerge in India only in the early 1990s. Following COBRA, six of the ten largest national firms in computers in terms of sales belonged to financially powerful economic groups, especially of the banking sector. SID, the biggest at the time of the National Informatics Policy of 1984, was controlled by Bradesco, the largest private bank; Itautec by Banco Itau, the second-largest (which became the largest by the end of the 1980s), as were nine out of ten suppliers of banking automation equipment.[12] SID was a creation of the Machline (Sharp) group, the largest consumer electronics group in Brazil. Several of these were also active in other segments of the electronics industry, notably SID and Itau. It is very significant that the banking sector is also the largest user of EDP equipment. This structure potentially allows intragroup planning of investments and intergroup coordination too if state policy is consolidated and/or coordinated effectively with the private sector, as in the Korean case. However, there are large numbers of small independent firms.[13]

(5) Following from the four characteristics above and also a cause of them, Brazil's institutional framework for policymaking and the instruments used to promote the industry have differed significantly from those of Korea and India, despite several commonalities. Even when the instruments used have been the same, their use has been different in important ways and the political and economic context different, giving rise to different results.

THE EVOLUTION OF BRAZILIAN ELECTRONICS POLICY AND THE POLICYMAKING INSTITUTIONAL FRAMEWORK

Brazil's policies toward the electronics complex of industries were not integrated at the level of policymaking institutions. Electronics strategy, if one can speak of one, emerged out of independent initiatives in the telecommunications, informatics, and, later, components segments. Toward consumer electronics there was no specific policy, its development being left to market forces within the ISI policy regime. When one talks of electronics strategy in Brazil, one is essentially talking about two independent initiatives in informatics and telecom from the late 1960s on.

Informatics

The First Policy Phase

In informatics, several factors converged to produce pressures for a new computer policy:[14] (1) Within the BNDE there was an increasing emphasis on greater vertical integration and diversification of the industrial structure. (2) Local computer development efforts to meet data-processing needs had led to the emergence of a group of experts who acted as a pressure group pushing for greater local development of the industry. (3) The Brazilian Navy's needs for modernization and independence from foreign technology for the electronic equipment of its British-supplied frigates led it to support local development. The "Guaranys Project," a joint Navy–BNDE initiative, created a Special Working Group (GTE 111) in 1971. The first important step was the creation of an interministerial agency, CAPRE (Commission for Coordination of Electronic Processing Activities), in April 1972.

CAPRE's mandate was the rationalization of computer purchases, installed base census, and human resource development. The Special Working Group, however, saw the opportunities for a local minicomputer industry from world technological trends. They recommended the establishment of a minicomputer industry based on a tri-pe (three-legged) joint venture between the government, a local firm, and a foreign firm. It was also decided to set up a prototype development project for a minicomputer at the University of São Paulo and a software project at the Catholic University of Rio.

These initiatives resulted in the creation of COBRA (Computadores Brasileira) in July 1974 to manufacture minicomputers. It was owned by the state-owned BNDE-funded Digibras, by Ferranti (UK), and by E. E. Equipamentos Eletronicos, a private firm. It was to manufacture the Ferranti-designed Argus 700 mini.[15] This was the high point of the first phase of informatics policy.

However, COBRA (and Argus) was a commercial failure. In December 1975, under the impact of the 1973–74 oil shock and balance of payments crisis, CAPRE was given additional powers to regulate all imports of computers. This power in the hands of CAPRE, combined with the existence of COBRA and the new Geisel regime's (dating from 1974) Second National Development Plan, which included "basic electronic industry" as a priority sector and in general pushed for backward vertical integration within the ISI framework, laid the basis for the Brazilian minicomputer industry.

However, the problem was that local capital was more interested in data-processing services and hence imports. In 1976, IBM do Brasil announced that it intended to start local manufacture of its System 32 mini. In July of the same year, CAPRE recommended that for national security reasons a domestic mini, micro, and peripherals industry be created "with total dominion and control of both technology and decision-making within the country."[16] The logical consequence of this was that IBM's request was turned down, and in June 1977 an open invitation was issued to nationally owned corporations to put forward proposals for the manufacture of minis. Nine firms entered the competition, two in conjunction with MNC's, the latter themselves submitting proposals for wholly owned ventures. CAPRE accepted three proposals from national firms, rejecting the MNC proposals.

This model, which came to be known as the National Model, had two features—namely, that only national companies could enter minis and micros, and each piece of foreign technology could be purchased only once. In January 1977, the Economic Development Council had established the criteria for approvals in the newly reserved industry. These were: extent of national ownership; export prospects; technology transfer; existing market structure and the impact of new entrants. These were the criteria used to select the three private companies (and of course, COBRA) in the competition for entry in 1977. These were SID, Labo, and EDISA, all of which were controlled by large industrial and banking groups interested in automating their operations. In December 1978, CAPRE also barred IBM and Burroughs from manufacturing small mainframes that might compete with minis.

The Second Phase: 1977–84

The second phase of Brazilian informatics policy—the creation of a reserved market for national firms in minis and micros—began in 1977. By 1983, there were over 100 Brazilian computer firms, which commanded 46% of gross sales of the industry, almost all in minis and micros. COBRA—the "national champion"—had lost its lead to private firms.

In December 1978, General Joao Figueiredo succeeded Geisel as president. Figueiredo was the chief of the SNI (National Intelligence Agency). A new committee, the Cotrim Committee, dominated by the intelligence community, submitted a report emphasizing the strategic importance of the informatics industry and recommending the abolition of CAPRE and the creation of a more powerful Special Secretariat for Informatics (SEI), directly under the National Security Council (CSN) rather than the SEPLAN. The Cotrim Report criticized CAPRE for not being nationalist enough on software and microelectronics dependency. SEI was duly formed in December 1979, under the CSN reporting directly to the President—an evolution of the National Model.

SEI eventually adopted an even more nationalist line in that it extended the Model to other segments of the electronics complex in the spirit of backward vertical and horizontal integration. Thus it created the Special Commission on

Software and Services in March 1980 to give strong preference to domestic EDP procurement and local software development. During the summer of 1980, a number of normative acts and directives were passed emphasizing local procurement, especially from national firms, by government agencies. SEI was restructured in 1981 in that an Advisory Council consisting of both private and public sector representatives was created, and its mandate was widened. This further strengthened the Model by making entry into reserved products manufacture even more subject to locally developed technology, software, and components. SEI's powers were further expanded to the approval of R&D projects in informatics in September 1981, and such approval made an "indispensable prerequisite" to import of equipment and tax incentives for R&D.[17]

During this 1979–84 second phase of policy, SEI also began to coordinate policy across segments to include microelectronic components, telematics, and transborder data flows (TBDF). Informatics policy began to overlap with telecommunications and components policies, which had hitherto developed independently from separate origins, to the extent that there were any specific policies at all for these segments. This happened just as it was becoming clear that computer and communication technologies were converging due to semiconductor-level developments.

The Third Phase: The National Informatics Policy of 1984

The third phase of Brazilian informatics policy began in 1984 with the passage of the National Informatics Policy as law on 29 October 1984.[18] Informatics was defined as comprising the entire microelectronics-based complex of industries, including computers, software and peripherals, telematics, TBDF, and so forth, including semiconductors and telecom. National corporations were defined as those headquartered in Brazil with decision-making, capital, and technological control in the hands of Brazilian residents or domiciles, or in the public sector. Direct state participation was restricted to cases of market failure; the state was only supposed to guide, coordinate and promote. The principal instrument of intervention remained the market reserve policy—that is, import control and MNC entry barriers extended for another eight years till 1992. However, some new institutional mechanisms of intervention were created: (1) CONIN (National Council on Informatics and Automation), consisting of 16 ministers and 8 other representatives, mostly of industry and science, headed by the Minister of Science and Technology, was created to orient informatics policy. (2) PLANIN (National Plan for Informatics and Automation) was instituted, with a three-year duration. Four matters were covered—use, production, R&D, and human resources. (3) Fiscal and financial incentives were given to national companies, especially for priority areas such as software and microelectronics. (4) Preference for national companies in state purchases was to be stepped up. (5) The

Informatics Technological Center Foundation was created under SEI to integrate R&D across companies, universities, and research institutes. (6) These changes were reinforced by the creation of the Ministry of Science and Technology (MCT) by the elected civilian government in 1985. SEI came to be subordinated to the MCT, as did other agencies such as CNPq. Software property rights and protection against infringement were also instituted. In sum, the new law and the changes that followed represented the integration of strategic policy institutions of the increasingly technologically convergent segments of the electronics industry.

Telecom Policy before 1974

Telecommunications policy in Brazil had originated and developed independently of, and earlier than, informatics policy.[19] The first major policy statement was the Brazilian Telecommunications Law of 1962. It called for a complete restructuring of the sector. The federal government was given sole authority over the development of all public telecom services, although it did not nationalize them. It established the Brazilian Telecommunications Company (EMBRATEL) in 1965 and also the National Telecommunications Fund. EMBRATEL is the national long-distance telephone company; since 1975, it has also been in charge of telex and datacom services. In 1967, the Minicom was created. The newly created CTB (Companhia Telefonica Brasileira), the Canadian-owned company that was responsible for 68% of the telephones in Brazil, was nationalized. In 1972, TELEBRAS was created and granted a monopoly over telecom services including over EMBRATEL. By 1973, TELEBRAS had merged the existing local companies into 25 regional companies, one for each state, and standardized tariffs. Two major internal sources of finance were EMBRATEL and the operating companies. The former was obliged to transfer 80% of its international service profits to the National Telecommunications Fund; the latter were allowed to earn 12% and over on capital. These two sources of finance made TELEBRAS largely self-supporting in its rapid expansion and modernization and allowed it to be a constant contributor to government revenues.

Three Objectives and Major Laws After 1974

TELEBRAS grew to become the eighteenth-largest company and second-largest absolute profit-maker in Brazil by 1982, despite tariff revision since 1974 consistently lagging behind inflation.[20] Thus the first objective of national telecom policy—that is, national control over the telecom service sector—was achieved and consolidated by 1974. However, telecom equipment manufacturing was still totally under MNC domination, and there was no significant local technological capacity in equipment or R&D effort. To gain control over telecom equipment manufacturing and to develop a national R&D capability were the second and third objectives of national telecom policy.

After 1974, the government moved on both these fronts with vigor. The strategy was to employ indirect intervention using TELEBRAS' monopoly over telecom services and hence monopsony power. The strategy also relied on in-house R&D capacity creation by TELEBRAS, backed by direct government support.

Three important laws were passed after 1974. The first, Law No. 102 of January 1975, defined the TELEBRAS philosophy of reducing dependence on foreign technology and increasing local R&D. It gave Minicom the authority to identify and deploy the available technological resources within the country. Minicom created GEICOM (Interministerial Executive Group for Components and Materials for Communications) in 1975, which was later to control telecom equipment and technology imports (of TELEBRAS, the monopsonistic purchaser) for saving foreign exchange. Law No. 102 led directly to the creation of TELEBRAS' R&D center, CPqD, in 1976.

The second main law, No. 661 of 15 August 1975, was the Industrial Policy for the Telecommunications Sector. This involved the use of the TELEBRAS monopsony to force the existing MNC equipment manufacturers to fall in line with national strategy. There were two aspects to this. One was that the MNCs were forced to dilute their holding of the voting shares of their Brazilian subsidiaries such that the majority of these shares became Brazilian-owned, even though the majority equity holding may still be foreign.

Law No. 661 also laid down the choice of technology to be developed by CPqD. A long-range decision to go digital was taken then.[21] However, in the short term the switchover was to be from electromechanical to SDS-SPC electronic technology for public-switching equipment. CPqD was to develop a family of digital (TDS) switches under its TROPICO program, and, most importantly, TELEBRAS was to insist on only this technology from all MNCs and to reserve 40%—later 50%—of its purchases for the CPqD-developed technology.

The third main Law, No. 622 of June 1978, called Policy for Technological Development and Acquisition of Equipment for the Telecom Sector, laid down a requirement that 51% of the voting shares be Brazilian-owned. The four major public switching equipment producers, Ericsson do Brazil, Nippon Electric, Standard Electric, and Equitel/Siemens eventually complied. A progressive transfer of the majority of voting shares to major Brazilian financial groups took place after the law. The idea was not to nationalize the companies, but to control their decision-making and hence their technology, and also to maintain competitive market structures.

These measures had the desired effects on the MNCs. The four major MNCs, led by Ericsson, began to comply with the government's demands. In order to gain market share in a situation of overcapacity and stiff competition, Ericsson negotiated a new agreement with its parent in Sweden and began to organize the large-scale transfer and local development of SDS technology. TELEBRAS used a large 800,000 line order for São Paulo to insist on transfer of the AXE–10 SDS system in 1982. The terms of the contract were far better than any before, covering exchange, transmission, and peripheral equipment, including full technical specifications and designs, thorough training, details of components and machine tool suppliers, and other vital information for finance, marketing, and purchasing, all for lower costs, patent and trademark rights, and without export restrictions.

In 1980, TELEBRAS abruptly dropped SDS-SPC as the favored technology and began insisting on TDS (digital) technology, canceling all orders except Ericsson's São Paulo contract. ITT sold off its shares in SESA and dropped out, leaving the latter all-Brazilian. However, Siemens/Equitel began to follow Ericsson's lead and adopt technology transfer as a competitive strategy to gain orders and market share. Equitel agreed to join a joint venture for TDS technology development, starting with smaller exchanges in collaboration with the largest national telecom firm, Elebra, and from January 1983 with CPqD! This surely represents a success story of indirect intervention using monopsony power to mold MNC behavior.

Law No. 622 also gave GEICOM the power to control and coordinate reduction of telecom equipment and component imports to aid the policy of national control. It defined "majority ownership" as a majority of the voting shares. The implementation of the law had the effect of raising the national share of telecom sales from 10% to 59% between 1978 and 1980.[22]

But this reflected not "an increasing indigenous competitiveness of technological capability but rather . . . a change in definition."[23] This led, after 1980, to controversy and a clash between Minicom and the more nationalistically minded SEI, as SEI was trying to extend its powers from informatics to telecom. SEI, in the areas under its jurisdiction, continued to insist that "national meant at least 70% if not 100% Brazilian ownership." However, despite this controversy, we have seen that the strategy as a whole has resulted in a major increase in local technological capacity and local manufacture by MNCs of SDS and then TDS systems, including technology transfer. Thus the strategy must be counted a success in promoting national capabilities.

Presidential Guidelines 1979

The Presidential Guidelines on Telecommunications of March 1979 took another important step in Article 5, which says: "The Ministry of Communications shall work jointly with the Department of Planning and the Ministers of Finance and of Industry and Commerce for the consolidation of the process of integration of the areas of telecom and informatics."[24] This gave formal recognition to the convergence of telecom and informatics, and with the creation of SEI began the process of institutional integration of policymaking for the two segments.

In January 1979, EMBRATEL was given the authority to operate the new datacommunication services, which could be public-access or private (leased facilities) in character. Private networks could be set up, using the facilities of the National Telecommunications Systems. Telematics policy in Brazil, governing the development of computer communications (or datacom), including international database access, came into existence only in 1980, after the emergence of such technologies and facilities. Policy in the field came about as a result of two directives: In May 1980 Minicom gave the pilot VIDEOTEX project to the telecom company of State of São Paulo, and in July 1980 SEI created the Special Commission on Teleinformatics.

The three top agencies, Minicom, CSN, and SEPLAN, acted through subordinate institutions—that is, TELEBRAS and its subsidiaries, including EMBRATEL, state companies, and RADIOBRAS; SEI; and CNqP and the Brazilian Institute of Information Science and Technology (IBICT). The institutional framework of telematics and TBDF policy consisted, therefore, of both Minicom and SEI, as well as IBICT and CNPq under SEPLAN. The two sets of Presidential Guidelines of Informatics and Telecommunications of 1979 also specified the need for the creation of a national datacom network and the creation of automated databases. Fundamental policy planks were that services be provided in a competitive environment, that national control over the creation and ownership of databases be promoted, and that the privacy of data users and producers should be protected. However, there was no massive expansion of a government-funded datacommunications infrastructure to provide demand for an ISI strategy in informatics and telecom equipment manufacture and local innovation on the scale of Korea's NAIS system.

Components Policy Evolution

Electronic components policy did not exist until the creation of SEI, being left to market forces, including export oriented MNC investments in the Manaus Free Export Zone. However, the National Informatics Law of 1984 defined microelectronics as part of informatics activities and included it in SEI's jurisdiction. Policy has been ISI-oriented, with national private firms being the market agents, in line with informatics policy. State-supported R&D has paralleled that in telecom and computers.

The main instrument of intervention was the market reserve policy. The main tools were import control through GEICOM, fiscal and financial incentives for production and R&D, and human resource development. The products chosen for national mastery were ASICS, in both semicustom and full custom device types, and in both bipolar and MOS technologies. This was a niche strategy, unlike Korea's Samsung's attempt to compete in standard memories. The main national firms were SID and Itau. There were many MNCs assembling discretes and ICs. The focus on production neglected Brazil's competitiveness potential in ASIC design while being doomed to inefficiency due to low volumes in a small domestic market.[25]

The Decline of the National Model and the Shift in the Role of the State

The 1984 policy regime ran into trade pressures from the United States under Section 301 of the US Trade and Competitiveness Act and, with the Collor regime's drastic opening up of the economy from early 1990, had to give up on much of the aims of the extended National Model of 1984.[26] Brazil had to open its markets to foreign software and extend software copyright protection in 1987. In 1989, the dispute was ended, with Brazil agreeing to several concessions, including clarification of its informatics rules, reduction of high tariffs on semiconductors to promote FDI, and agreement to discuss its regulation of FDI, technology import, and market access generally. In a way, the very success of the National Model in creating the developing world's largest computer market attracted pressures toward the dismantling of the Model. Sections of domestic industry outside electronics played a role in the dismantling of the National Model in the late 1980s, in alliance with sections within the state apparatus, including Minicom, catalyzed by US pressure, with the resistance of domestic hardware producers declining. SEI was abolished in 1989, and informatics, including teleinformatics and all microelectronics-based industries, was brought under the Department of Informatics and Automation Policy (DEPIN) in the Secretariat for Science and Technology in the President's Office.

However, from the late 1980s on, the role of the state began to shift from direct production of the COBRA type to infrastructural, human resource, and R&D support to private enterprise to aid indigenous technology development and competitiveness, the best example being the attempted development of an indigenous 32-bit superminicomputer in the late 1980s.[27] However, Brazilian firms were not as large and competitive as Korean firms, and, for that reason, they did not have the financial and technological resources or the capacity to license foreign technology on the scale of, say, Korea's Samsung. Furthermore, SEI's CTI was not doing R&D on the scale of Korea's ETRI, nor did Minicom have the procurement leverage, the power, or the inclination to create a demand-generating datacom infrastructure like that of Korea in the 1980s. Indeed, Minicom was prone to clientelist politics and in May 1984 relaxed the reservation of large-scale switching equipment procurement for CPqD's TROPICO RA (up to 20,000 lines) main exchange and went in for large-scale procurement of medium and large exchanges (4,000–10,000 lines) on grounds of delay in supplying pent-up demand.[28] Therefore neither the state nor the private sector was able to carry out a policy shift of the kind in Korean electronics' third and emerging fourth phases.

Brazil's Electronics Policy Trajectory: A Summing-Up

The overall trajectory of Brazilian policy and development was as follows:

(1) First there was a thrust to acquire national control over ownership, decision-making, and technology in the computer and telecom industries during the

1970s from the entrenched MNCs using nationalization of telecom services, creation of a "national champion" in COBRA, the market reserve policy, in small computers and GEICOM's import controls as instruments.

(2) Technology transfer and local technology development were promoted using TELEBRAS's monopsony, CPqD's R&D and telecom network expansion to create a national market as policy instruments.

(3) The early 1980s saw the institutional consolidation of informatics policy in SEI, followed by the rapid expansion of the National Model, culminating in the 1984 policy that marked the closer integration of policy and policy institutions for all microelectronics-based industries and deeper thrusts in superminis, software, semiconductor, and enhanced tele- and datacom services. However, continued ISI did not lead to competitiveness due to lack of selectivity in ISI, scale economies, and demand-generating complementary investments, although considerable technological capability and the largest developing-country electronics market had been created.

(4) From the late 1980s on, the partial dismantling of the National Model took place partly as a result of US pressure and partly as a consequence of general deregulation. The state's role, as in Korea, tended to shift from regulation and state ownership to support for private sector initiatives with infrastructural and R&D backup. However, the industrial base has been laid in terms of technology, human resources, and infrastructure for possible future competitiveness in several niches, given appropriate international alliances and an appropriately supportive state.

BRAZIL'S ELECTRONICS STRATEGY: QUESTIONS THAT NEED A POLITICAL EXPLANATION

Four Broad Questions

The course of Brazilian electronics policy and development raises four interrelated questions, which require a political explanation in that they require an understanding of the nature of state power in Brazil.

(1) Why did Brazil, with a strong, well-entrenched MNC presence in the industry, move to strengthen local production and national firms in informatics and telecom, especially after 1974? This becomes even more of a puzzle if one considers that during the "miracle" years of the late 1960s and early 1970s MNC investment was being encouraged, and triangular tie-ups were actively promoted. How did it politically manage to control MNCs in informatics and telecom?

(2) Why did Brazil choose ISI in informatics and telecom—the industrial equipment segments—rather than assembled consumer electronics or component exports, when it could very possibly have found a niche in the booming world electronics trade in the latter segments?

(3) Given an ISI strategy, why did it choose private national firms and MNC subsidiaries as the market agents rather than state-owned enterprises when the latter were the chosen market agents for ISI in heavy and capital goods industries in the 1970s?

(4) Why did it take until as late as 1984 to integrate the various policymaking and implementation agencies at an institutional level if the government was serious about promoting electronics on an ISI basis? Why were only limited instruments such as import control used for the most part? Why did the government not plan for the electronics complex as a whole rather than in a piecemeal, segmented manner, leaving so much to the market?

What were the political and institutional constraints and pressures that shaped the contours and trajectory of policy and institutions in these directions and ways? To understand this, one has to be able to specify the relationship of the state to the principal economic and political forces, those of foreign capital and domestic capital, taking into account the fact that the military was in power from 1964 until 1984. In what follows, we attempt an interpretative political economy analysis of the evolution of Brazilian electronics policy from the foregoing survey and the literature on Brazil.

1. Why Did Brazil Move to Create an Autonomous and Indigenous Electronics Industry?

The answer to the first question has been given at great length and in detail by Peter Evans and Emmanuel Adler.[29] The first thing that stands out about the drive to develop a nationally owned and controlled electronics base is the dominance of national security concerns and ideology as motivating factors on the part of the military regimes from the late 1960s on. Later, after 1974, balance of payments pressures and the start of a new round of ISI in the Second Plan played an important role in the drive to initiate local production.[30] However, unlike in Korea, the economic motivation remained subsidiary.

The genesis of Brazil's informatics policy lay in the Navy's acquisition of British frigates in the late 1960s. The Navy did not want to be permanently dependent on foreign (British Ferranti) computer and electronic control systems. It promoted a project to develop a domestic computer for the purpose, resulting in CAPRE and COBRA. The motivating force behind this development was national security. It was felt that a manufacturing and technological foothold had to be created in the field.

The other major source of Brazil's autonomy drive in computers, simultaneous and meshing with national security, was an ideology of antidependency. Adler terms this "pragmatic antidependency," because it did not seek to de-link itself from world capitalism, as advocated and thought necessary by *dependencistas*; rather, it sought to mobilize state and private institutional resources to bargain for better terms from MNCs. It meant not nationalizing foreign firms but limiting their domination, or gradually rolling back their control (as in telecom), of the domestic market. It also meant negotiating for better terms of technology transfer and local manufacture. The principal instruments used were import control, entry barriers (the market reserve in smaller computers), and state-supported R&D. The pragmatic antidependency approach (with emphasis on the first word) explains the autonomy drive per se as well as its step-by-step, nonconfrontational implementation.

The approach succeeded in small computers, since at the time of instituting the market reserve there were no MNCs involved in their local manufacture. Evans emphasizes this point, arguing that had IBM or DEC been interested in manufacturing their minis in Brazil in the early 1970s, they would have faced no opposition. It was when IBM tried to initiate such manufacture in 1976 after the nationalist policy had already been formulated that it was not allowed. And, conversely, had IBM been locally established in minis in 1977, the market reserve policy would not have got off the ground unless the Brazilian state was willing to risk a major confrontation with foreign capital in general.

In the telecom case, where there were established MNCs, the approach was one of molding their behavior by gradually stepping up pressure while simultaneously strengthening the state's hand by institutional changes such as nationalization of services and the creation of CPqD. These "pragmatic" antidependency steps were taken against the backdrop of continuing openness to FDI in other sectors of industry, including tri-pe-style ventures involving the state enterprises.

The economic rationale—saving foreign exchange—for nationalist electronics policies emerged as part of a new round of ISI after the 1973 oil shock and the balance of payments crisis. The Second National Development Plan launched by the new Geisel regime in 1974 consisted of a concerted thrust into heavy and capital goods industries, led by state enterprises. The Plan defined "basic electronics industry," including computers, as a priority, seeing this as a growth area. It also envisaged a major thrust toward an indigenous R&D capability for technological autonomy for both national security and industrial competitiveness. Thus national security, ideological antidependency, and the new ISI and balance of payments management strategy all combined by the mid-1970s to generate nationalist computer and telecom policies, albeit in different state agencies and by different groups of people, each acting independently.

The important point here, made by both Evans and Adler, is that the national security and pragmatic antidependency ideologies informing computer policy—and later all electronics policy—arose and came to power within the institutions of the state apparatus. It was these ideologically motivated individuals and groups in control of policy institutions who were responsible for the electronics promotion strategies, their institutional innovations, and the specific use of instruments from the early 1970s through to the early 1980s, at least. Domestic capital in the industry became a factor influencing policy only after the creation of protection-dependent local firms in computers after 1978.

2. Why Industrial Electronics ISI and Not Labor-Intensive Exports?

The reason for the industrial equipment segments—computers and telecom equipment—to be singled out for promotion would appear to follow from the factors behind the indigenization and autonomy drive. These were the segments crucial for self-sufficiency in defense, heavy industry, infrastructure, and high technology. A decision at that time to develop an indigenous computer industry would have to be within an ISI framework. Though it was possible to develop an indigenous minicomputer industry and to gain control over MNC telecom equipment producers, it was not possible for any developing country to become competitive in such areas in the mid-1970s.

The question, however, can be raised as to why Brazil did not go in simultaneously for a labor-intensive consumer electronics and semiconductor assembly industry, taking advantage of its low wages and attractiveness to MNCs? Such a strategy would provide important forward linkages for an indigenous components and parts industry. And Brazil, it could be argued, was in a position to put pressure on MNCs to export from Brazil as a condition for access to its substantial domestic market. Such an export-oriented strategy would have neatly complemented ISI in industrial electronics and also have served the objective of shoring up the balance of payments.

Brazil did, in fact, try to encourage exports of consumer products and components by MNCs in the Manaus Free Export Zone, set up in 1967. IBM and Burroughs were the principal exporters of electronics from Brazil into the early 1980s. However, such exports remained very small compared to Korea and the other East Asian NICs. It is not quite clear why this was so—whether Brazil's wage levels were uncompetitive compared to those in the Far East, whether MNCs were by then tied to already sunk costs elsewhere, or whether there were other reasons. It would seem that there was a lack of policy initiative as well as a lack of economic incentive for MNCs to do so. No "pragmatic antidependency guerrillas" or any other policy initiators arose for a components or consumer electronics exports strategy. This only underlines the lack of economic–institutional consolidation in electronics policy in Brazil until 1979 or even 1984. Also, MNC-owned intrafirm export-oriented assembly operations with little technology transfer might go against the grain for the backward vertical integration thrust of post-1974 ISI as well as against the ideological grain of the pragmatic antidependency perspective.

3. Choice of Market Agents:
Three Possible Courses vis-à-vis MNCs

The large role allowed MNC subsidiaries in Brazilian ISI and the relative absence of state enterprise, in the midst of a national security and antidependency-motivated ISI drive, requires explanation, especially as state enterprise was prominent in heavy industry. Excluding MNCs would have meant either (1) nationalizing them or (2) forcing them to dilute their foreign ownership or (3) limiting their entry into specified product categories. The first course would have been highly confrontational, and Brazil followed a pragmatic policy on MNCs, avoiding unnecessary confrontation—or, to put it differently, the most confrontational course politically feasible. In electronics policy it followed each of the above three courses, but only where it was safe politically. The cases below illustrate the highly constrained nature of the Brazilian state's options regarding MNCs.

Nationalization would have been detrimental to the economy in the short term, given the large role played by foreign firms in several key industries. It would also alienate important sections of domestic capital allied to MNCs, not to speak of the military. Therefore, for antidependency advocates to push for such a course could have been politically suicidal for the entire autonomy drive. The Brazilian state avoided such nationalization, except in the case of telecom services—a segment that was state-owned almost everywhere and would therefore not have been seen as a threatening precursor to further moves against MNCs.

The second course, of forcing dilution of foreign control, was followed in the case of the telecom equipment producers from 1978 on. However, here, too, the state was pragmatic and forced dilution in a watered-down way. It forced MNCs to surrender to domestic (private) control the majority of only the voting shares, leaving them still largely foreign-owned, though nationally controlled. This was clearly a compromise solution that diluted the concept of the national firm and led to a controversy between Minicom and the more nationalist SEI, which was implementing a strictly national firm policy in computers.

The third course—that of limiting foreign firm entry into specified product lines—was the policy followed in the case of mini- and microcomputers. The market reserve policy did not curtail the existing activities of the established MNCs in the industry. It merely reserved for domestic firms a category of products that were not then being produced in the country by MNCs. The political battles that were fought within the state apparatus by the "ideological guerrillas" of antidependency, and those between them and the MNCs, have been described by Evans and Adler. Both place great emphasis on the fact that the MNCs were not already present in the reserved products, for the political success of the policy.

For this reason it would not have been possible to carry out such a market reserve strategy in telecom where MNCs were virtually monopolizing the industry. Nor would it have been possible to do so in consumer electronics, since here, too, MNCs were present in the main product lines—radio and television—aimed at the domestic market. Thus, a Korean-style strategy of building up local firms as assemblers of consumer products would have been ruled out as politically unfeasible, if ever it was sought to be attempted—not that Brazilian electronics policy ever gave consumer electronics a pivotal role either as an export item or as a source of domestic demand for a fledgling component industry.

Another plausible explanation of such pragmatism is that Brazil did not have to go all the way with ISI for autonomy out of national security concerns. It did not need to indigenize components production and technology. It was sufficient to develop a capacity to design and manufacture computers and telecom equipment indigenously using imported components and peripherals and to build up toward system capability. It did not require immediate self-sufficiency as a military necessity, because as a Western-leaning nonaligned country in which MNCs had a large stake, and not faced with any immediate security threat, it did not face the threat of cutoff of electronic equipment and components.

What also needs explanation is the fact that Brazil did not employ state-owned enterprises as its chosen market agents to create an autonomous electronics industry. In the electronics industry, the only state enterprise in production was COBRA, which was not sheltered from competition by government purchase preferences and eventually lost its lead. TELEBRAS was a giant, self-financing state enterprise, but it did not enter equipment production. The creation of productive state enterprises instead of nationalization or equity dilution would have avoided confrontation with foreign capital.

It is necessary to examine here the motivation behind state enterprise in the context of the post-oil crisis round of ISI in heavy and advanced technology industries. State enterprise burgeoned under the right-wing military regimes of 1964–84, which were officially pro-private and pro-foreign enterprise. In mining and manufacturing 14 new enterprises were created in the period between 1960 and 1969, in addition to the 11 in existence; 24 were added during 1970–75 and 16 during 1976–80. In transport and communications, electricity and finance—all key infrastructural sectors—1960–69 saw 16 added to the existing 33, 1970–75 saw 36, 1976–80 another 12. In the "other" category, 9 were added to an existing 46 during 1960–69, 28 during 1970–75, and 5 during 1976–80. By 1979, there were 56 state firms in manufacturing alone. The new informatics and telecom policies thus coincided with the proliferation of state enterprise.

Thomas Trebat lists the following six competing hypotheses to explain the Brazilian state's intervention in the sphere of production:[31] (1) the weak private sector hypothesis—where private enterprise lacked capital, technology, and managerial talent; (2) the economies-of-scale hypothesis—also called "natural monopolies" or "optimal size of firm" hypothesis—which predicts public enterprise in those sectors where scale is important and indivisibilities may be present, such as energy, telecommunications, and other classic infrastructure; (3) the ex-ternal economies hypothesis, which predicts public enterprise in the public goods sectors (roads, bridges), where private owners are not rewarded for the external technological or pecuniary economies generated; (4) the dynamic public managers hypothesis—public enterprises in developing countries are able to recruit managers with superior abilities, who take advantage of the growth opportunities deriving from the scale and complexity of the firm's operations; (5) the natural resource rents hypothesis—ownership of natural resources generates unearned incomes or rents, public appropriation of which is an important economic motivation for resource nationalization; and (6) a political–historical hypothesis, which allows that societal (decision-makers' ideological) preferences for public over private and domestic over foreign ownership may lead to the creation of public enterprises. Trebat argues that public enterprises created for political– historical reasons will probably be randomly distributed across sectors.

Except for the last two hypotheses, the others derive state behavior from the usual justification for public involvement in production, based on market failures due to the lumpiness and long gestation (high uncertainty) of the investments required and externalities. Only the last allows some role for ideologies and the working of anti-foreign private interests on state policy.

Werner Baer takes a more political view, dismissing as inadequate the view that state intervention occurred merely to correct for market failures. While allowing for ideological factors, he emphasizes the incoherent and ad hoc nature of state intervention. But he also attacks the notion that the state represented domestic against foreign capital, and the view that it was controlled by an alliance of the two and acted at their behest to provide subsidized infrastructure or that the state was autonomous from class forces. These views do not explain the "aggressive expansion of state enterprises at the expense of the private sector."[32] The debate over statism in the 1970s was evidence of the tensions between private capital and the state over the latter's expansion.

One may draw the conclusion, therefore, that the best explanation of the expansion of state enterprise in different sectors of the economy is as the *result of complex power struggles over interests and ideologies* within the state apparatus, with different factions linked in complex criss-crossing relationships with private firms and domestic capital. But it does not support Trebat's hypothesis that political–historical factors, if any, would result in state enterprise randomly distributed across sectors. Rather, a political power struggle explanation would lead one to expect state enterprise in those sectors where it would not come into conflict with entrenched private capitalist interests, domestic or foreign. A random distribution hypothesis assumes near-total autonomy on the part of the state's decision-makers and their ability to ride roughshod over the interests of private capital. One must also take a much more qualified view of Adler's explanatory emphasis on ideology as the driving force behind the National Model in computer policy. This view also assumes far too much autonomy on the part of policy-makers. We have seen from the contrasting cases of telecom and MNC-dominated consumer electronics and components that ideology alone cannot explain the autonomy drive, still less the strategy employed.

If one accepts such an explanation of Brazilian state enterprise creation, then the nonchoice of state enterprise as the market agent in electronics (other than COBRA) falls into place:

(1) The only act of nationalization that took place—that of telecom services in 1967—was done against a Canadian company that was not a major MNC and not from a politically powerful state.

(2) The creation of COBRA in 1974 part-owned by a state holding company Digibras fell into the tri-pe pattern then being promoted. State entrepreneurship here did not clash with MNCs or domestic capital, since there were none in minicomputers. Minis were an "empty space," and the explanation of state enterprise being due to the weakness of the private sector could also apply. Hence, there was no basis for political opposition to COBRA, while there were strong "push" factors within the state apparatus emanating from the national security and pragmatic antidependency lobbies, obviating the need for "pragmatism."

In telecom equipment, however, very considerable pragmatism was necessary, as there were four established MNCs. Not only was there no "empty space," there was no possibility of local capital being technologically strong enough to enter the field. In consumer electronics and components there were again no "empty spaces." MNC subsidiaries dominated these markets. However, private capital could (and did) effect entry if it wanted to. Unlike in the case of telecom equipment, there was no way out via the exercise of monopsony power in these segments. This could probably explain the lack of policy initiatives leave alone state enterprise in these areas.

4. Why Was There No Integration of Policy Planning Across Segments?

The lack of institutional integration for overall strategic planning for the electronics complex until after 1979 at the earliest becomes explicable if one considers the institutional origins of the antidependency policies—or lack of them—in the different electronics segments. The primary purpose that would have been served by policy-institutional integration would have been the ability to forecast demand and coordinate investments (state, state-owned bank, and private) in the interlinked segments and, in turn, coordinate policies on R&D, MNC regulation, and technology transfer.

In the telecom sector, the initiatives that began with the Brazilian Telecommunications Law of 1962—the creation of Minicom and nationalization of telecom services in 1967 and the creation of TELEBRAS in 1972—were driven fundamentally by the felt needs of national security and a perceived economic need to establish a nationwide telecom infrastructure in a vast, diverse country.

TELEBRAS was necessary in its centralized form to be able to finance the expansion of services nationwide, including cross-finance unprofitable services necessary for national security from profitable urban and inter-city traffic. Economies of scale could be reaped only by a national holding company. The earlier, fragmented, foreign-owned system was incapable of expansion and modernization and uninterested in expanding services to remote, unprofitable regions that the military regime considered necessary for national security.

The second post-1974 objective was to establish a production and technological base in equipment manufacture. Here the state was greatly constrained by the presence of MNCs and could only move "pragmatically" using import control and procurement policy as tools. Until 1974, the main policy agencies, Minicom and TELEBRAS, had been inspired by national security and national integration ideologies, more than by antidependency; there would have been no perceived need to coordinate with, let alone institutionally integrate, Minicom with policy agencies for other segments, since they did not exist.

In informatics, the first policy agency, CAPRE, under SEPLAN, came into being in 1972. CAPRE's technocrats saw the possibilities of creating a domestic minicomputer industry. However, this group was quite distinct from the people who ran Minicom and TELEBRAS. Secondly, telecom and computer technology had not yet begun to merge on the basis of standard chips, modules, and networking possibilities. Thirdly, the informatics technocrats faced a different market situation and a different group of MNCs. Had they been dealing with the same companies, or had they been following the same goals and the same strategy of national ownership of the equipment industry rather than creation of a new industry by market reserve, there could very possibly have been interinstitutional coordination. Until 1979, the different electronics segments were therefore treated as independent industries under the jurisdiction of different ministries.

In consumer electronics there was no specific policy. This is not altogether surprising, given that MNCs dominated the industry and that the main relevance of the industry in the electronics complex as a whole was as a major consumer of simpler semiconductor components. As Brazil did not attempt national–autonomist ISI in components until the mid-1980s, components development was left to market forces or imports. Thus there was no need to plan for the provision of scale economies to domestic component producers, which would have been the function of the consumer segment.

The reason why Brazil did not attempt an across-the-board ISI strategy in electronics but limited itself pragmatically to any opportunities that opened up in informatics and computers can probably best be explained by the absence of any major military threat or the threat of cutoffs of supplies of vital electronic equipment and components for geopolitical reasons, Brazil being a pro-Western regime. Hence there was no need for a self-sufficency drive to supply the requirements of a domestic defense production or nuclear energy program (even though Brazil did initiate an ambitious arms industry and a nuclear program in the mid-1970s).[33] And, since the technological merger of telecom and computers based on developments in chip technology and systems architecture did not arrive until the 1980s, there was no pressing need to integrate planning for the electronics complex as a whole.

A final important issue is the question of why Brazilian policy appears to have used only some of the instruments available to it to promote the industry. In the

Korean case, we have seen that not only was import control used as a tool in an industry-biased manner, but selective credit allocation to firms by government-controlled banks was a key instrument. So was the discretionary use of governmental authority to regulate the terms of foreign equity holding in, and technology transfer to, subsidiaries and joint ventures. In the 1970s, Korea's EPB combined the power to regulate credit (and hence domestic private investment), foreign investment, joint ventures, and technology transfer in one decision-making center. All these closely interrelated factors were therefore taken into account in approvals. Later there was close coordination between the other key agencies. Korean legislation on FDI, technology transfer, and foreign borrowing by firms as well as credit allocation by national banks was framed in a way general enough to give state policymakers very great discretion in interpretation and implementation. This made it possible to craft the best possible promotion package for every approved investment from the point of view of competitiveness.

In the Brazilian case, the lack of institutional integration of policymaking agencies and hence lack of coordination of approvals of interrelated decisions led to the industry's development being haphazardly driven by domestic ISI-distorted market forces. A coordinated strategy could not be carried out because, apart from the complication of the MNC presence across the board, the power to approve foreign investments, joint ventures (equity and technology transfer), credit allocation, and import policy was not integrated in one agency. Industrial credit policy, even of government-owned banks, was not integrated with market reserve and selective entry approval. This was so especially because a substantial proportion of the banking sector was private and part of industrial/financial groups, and it followed distorted domestic market signals. Thus, even in minicomputers, where selective credit allocation, foreign collaboration and entry approval, and choice of the most promising firms as market agents *à la* Korea might have been possible, what happened through the 1980s was the proliferation of suboptimally sized mini- and microcomputer assemblers. Even in telecom there were far too many firms for a market of only a little over half a million lines annually for most of the 1980s, which was the international minimum efficient scale[34]—and this despite the fact that all Brazilian electronics legislation was, like Korea's, worded in a sufficiently general way as to give the executive agencies very considerable discretionary powers in interpretation and implementation.

CONCLUSION: IMPLICATIONS FOR STRATEGIC CAPACITY

One can conclude that the failure to integrate policy across the different segments of electronics and across the key decision areas like import control, credit allocation, local firm entry, and foreign investment approvals was substantially responsible for the failure to exploit such powers to create internationally competitive firms. This, in turn, implies a lack of strategic capacity, both in Deyo's limited original sense of political closure and economic institutional consolidation and in our extended sense of path-dependent policy flexibility. The massive prior presence of MNCs and their tri-pe alliances precluded political closure. Secondly, state institutions were never consolidated and could not effectively coordinate even after 1984, for precisely the same reason—penetration by a variety of interest groups. The presence of MNCs in alliance with domestic firms also precluded the mobilization of domestic peak private sector organizations as agents for policy implementation on the lines of the *chaebol* groups. As Evans and Tigre have argued, Minicom was characterized after 1984 by clientelist politics that often frustrated the nationalist initiatives of SEI and limited the scope of coordinated action on policy issues.[35] Thus Minicom's procurement leverage and its control over TELEBRAS' financial and technological resources, which SEI lacked, were never used to support a coordinated computer-communications strategy; nor was there any massive datacom infrastructure planned for such purposes, as it was in Korea.

Strategic capacity in Deyo's sense and policy flexibility and coordination were thus constrained by preexisting and path-created vested interests, effectively preventing the exploitation of the industrial, technological, financial, and policy assets that Brazil came to have by the early 1980s. Policy flexibility and interagency coordination in the shift in the state's role from regulator to facilitator was probably also constrained during the 1980s by the path-created ideological legacy of antidependency that defined it as ISI in equipment and technology. This probably constrained or even screened out any possibility of a simultaneous export effort in alliance with MNCs in labor-intensive assembly industries in the 1970s and selective promotion of appropriate international alliances with MNCs for accessing technology and markets from the late 1980s on.

NOTES

1. All figures in this section are from the following sources, unless otherwise mentioned: Elsevier Advanced Technology, Oxford, UK, Benn Electronics Publications, *Yearbook of World Electronics Data* (London: Elsevier), various years; Clelia Piragibe, "Policies Towards the Electronics Complex in Brazil," Texto No. 28, Centro de Estudos em Politica Cientifica e Tecnologica (CPCT), CNPq, 1987.

2. Parthasarathi, "Electronics in Developing Countries: Issues in the Transfer and Development of Technology" (Geneva: United Nations Conference on Trade and Development, 1978) Table 1.3, pp. 9–10.

3. Frischtak, "The Information Sector," Table 1, pp. 12–13; Emmanuel Adler, "Ideological Guerrillas and the Quest for Technological Autonomy: Brazil's Domestic Computer Industry," *International Organization*, 40, No. 3 (1986), 673–705, Table 1, p. 679; Piragibe, "Policies Towards the Electronics Complex," Figs. V and VI.

4. Frischtak, "The Information Sector," Table 1, p. 12.

5. Michael Hobday, "The Brazilian Telecommunication Industry: Accumulation of Microelectronic Technology in the Manufacturing and Service Sectors," UNIDO/IS.511 (25 January 1985), Table 1, p. 9; and calculated from Cae-one Kim, Young Kon Kim, and Chang-bun Yoon, "Korean Telecommunications Development: Achievements and Cautionary Lessons," *World Development*, 20, No. 12 (1992): 1831, Table 1.

6. Ibid., p. 9.

7. Claes Brundenius and Bo Goransson, "Technology Policies in Developing Countries: The Case of Telecommunications in Brazil and India," *Viertelsjahresberichte*, No. 103 (March 1986): 43–64, especially 50, 54–55.

8. Claudio Frischtak, "Specialization, Technical Change and Competitiveness in the Brazilian Electronics Industry," OECD Development Centre Technical Paper No. 27 (October 1990): 35.

9. Hobday, "Brazilian Telecommunications"; Wajnberg, "Brazilian Microelectronics."

10. This section draws heavily on the accounts of Brazilian scientific and technological development in Emmanuel Adler, *The Power of Ideology: The Quest for Technological Autonomy in Argentina and Brazil* (Berkeley, CA: University of California Press, 1987); Brundenius and Goransson, "Technology Policies"; Wajnberg, "The Brazilian Microelectronics"; Michael Hobday, *Telecommunications in Developing Countries: The Challenge from Brazil* (London and New York: Routledge, 1990); Hobday, "The Brazilian Telecommunication"; Hobday, "Telecommunications: A 'Leading Edge' in the Accumulation of Digital Technology? Evidence from the Case of Brazil," *Viertelsjahresberichte*, No. 103 (March 1986); United Nations Center for Transnational Corporations, *Transnational Corporations in the International Semiconductor Industry* (New York: United Nations, 1983); Fabio Stefano Erber, "The Development of the Electronics Complex and Government Policies in Brazil," *World Development*, 13, no. 3 (1985): 293–309; Piragibe, "Policies Towards the Electronics Complex"; Francisco Colman Sercovich, "Brazil," *World Development*, Special Issue on Exports of Technology by Newly-Industrializing Countries, 12, nos. 5/6 (1984): 575–599.

11. See Claudio Frischtak, "Specialization, Technical Change and Competitiveness in the Brazilian Electronics Industry," OECD Development Centre, Technical Paper No. 27 (Paris: OECD, 1990), pp. 13–18 and 51–52.

12. Frischtak, "Specialization," p. 29.

13. For accounts of the private sector in Brazilian electronics, see Frischtak, "The Information Sector," p. 16; Peter B. Evans, "State, Capital and the Transformation of Dependence: The Brazilian Computer Case," *World Development*, 14, no. 7 (1986): 791–808, and Table 4, p. 802; Evans and Tigre, "Going Beyond Clones"; United Nations Center for Transnational Corporations, *Transnational Corporations*, Annex I, Table III–10, pp. 211–213.

14. This section draws heavily on the accounts in Erber, "The Development of the Electronics"; Evans, "State, Capital"; Adler, "Ideological Guerrillas"; Piragibe, "Policies Towards the Electronics Complex"; Paulo Bastos Tigre, *Technology and Competition in the Brazilian Computer Industry* (New York; St. Martin's Press, 1983); United Nations Center for Transnational Corporations, *Transborder Data Flows and Brazil* (New York: United Nations, 1983).

15. Ferranti held only 3% of equity and agreed to transfer technology (Adler, "Ideological Guerrillas," p. 689).

16. United Nations Center for Transnational Corporations, *Transborder*, Annex II, B4, p. 307.

17. Ibid., p. 352.

18. For a brief account of the law, see Piragibe, "Policies Towards the Electronics Complex," pp. 3–5.

19. This section draws heavily on Hobday, *Telecommunications*; Hobday, "Telecommunications: A 'Leading Edge'"; Hobday, "The Brazilian Telecommunications"; Brundenius and Goransson, "Technology Policies."

20. Hobday, "The Brazilian Telecommunication," p. 10.

21. The Indian engineer Sam Pitroda, then US-based, was intimately involved in encouraging CPqD to go in for indigenously designed digital exchanges from the time of his three week visit to Brazil in the winter of 1975–76. This was based on his US experience, and from 1984 on he pioneered the same effort in India. Interview with Sam Pitroda, New Delhi, 20 May 1995.

22. Brundenius and Goransson, "Technology Policies," p. 55.

23. Brundenius and Goransson, "Technology Policies," p. 54.

24. United Nations Center for Transnational Corporations, *Transborder*, Annex II, A2, pp. 263 and 265.

25. Frischtak, "Specialization," pp. 19–26.

26. See Maria Ines Bastos, "How International Sanctions Worked: Domestic and Foreign Political Constraints on the Brazilian Informatics Policy," *Journal of Development Studies*, 30, no. 2 (January 1994): 380–404, especially 390–395.

27. For details, see Evans and Tigre, "Going Beyond Clones."

28. Frischtak "Specialization," p. 47; Evans and Tigre, "Going Beyond Clones," p. 1761.

29. The following account draws heavily on Evans, "State, Capital"; and Adler, "Ideological Guerrillas," and *The Power of Ideology*.

30. See the accounts in Jose Serra, "Three Mistaken Theses Regarding the Connection between Industrialization and Authoritarian regimes," in David Collier (ed.), *The New Authoritarianism in Latin America*, pp. 99–164 (Princeton, NJ: Princeton University Press, 1977); Thomas J. Trebat, *Brazil's State-Owned Enterprises: A Case Study of the State as Entrepreneur* (New York: Cambridge University Press, 1983).

31. Trebat, *Brazil's State-Owned Enterprises*, p. 37, Table 3.1, p. 38, Table 3.2, and pp. 32–35.

32. Werner Baer, "Political Determinants of Development," in Robert Wesson (ed.), *Politics, Policies and Economic Development in Latin America* (Stanford, CA: Hoover Institution, 1984), p. 69.

33. For a discussion of Brazil's military industrial complex, including electronics, see Helena Tuomi and Raimo Vayrynen, *Transnational Corporations, Armaments and Development* (New York: St. Martin's Press, 1984), chapter on Brazil.

34. Frischtak "Specialization," pp. 49 and 51–54.

35. Evans and Tigre, "Going Beyond Clones"; Frischtak, "Specialization," p. 47.

5

The Development of the Electronics Industry in India

India has sought to promote the electronics industry since the early 1970s. At that time, India was on a par with or even ahead of Korea and even Brazil in terms of industrial base and technological depth and had a head start in terms of institutions for integrated policy planning. However, these advantages have not been translated into comparable performance in output, exports and even broadbased technological depth. The reasons for India's failure to make effective policy take one into the realm of comparative political economy, for which an account of Indian electronics development is a necessary background.

INDIAN ELECTRONICS DEVELOPMENT: THE EMPIRICAL RECORD

Historical Background

Electronics production in India began with diodes in the early 1950s and became significant in volume in the 1960s with the production of radio receivers, led by foreign firms. Two public sector enterprises—Indian Telephone Industries (ITI), set up in 1948, and BEL, set up in 1954, under the Ministries of Communications and Defense, respectively—dominated electronics production. One of the government's most important policy initiatives, right from the outset, was the creation of "national champion"-type public sector electronics firms to pioneer local production. Policy-institutional consolidation was achieved early with the creation of the Electronics Commission (EC) and the Department of

Electronics (DOE) in 1970–71. The focus was on the industrial electronics segments, comprising, in the Indian classification, communications and broadcasting, computers, control instrumentation and industrial electronics (CIIE), aerospace/defense (now strategic electronics), and components. The Electronics Corporation of India (ECIL) was created in 1967 under the Department of Atomic Energy (DAE) to produce nuclear electronic instrumentation, but it subsequently became the "national champion" in mini- and large computers in the 1970s and 1980s, respectively, as well as diversifying into television and general instrumentation. ITI, the telecom equipment "national champion," produces switching, transmission, and peripheral electronic equipment. BEL, the defense electronics "national champion," produces specialized equipment and components for the defense sector. Computer Maintenance Corporation (CMC), which began in 1976 as the maintenance and services monopolist for IBM's installed base after the latter quit India in 1978, became a leading software and systems engineering firm. Semiconductor Complex Limited (SCL), the IC design and fabrication "national champion," became operational in 1984. The Electronics Trade and Technology Development Corporation (ET&T) was originally intended to be a bulk purchaser of components and materials. The "national champion" character of ITI, ECIL, CMC, and ET&T eventually became eroded with the freer entry of private and foreign firms into hitherto public sector preserves.

Between the 1960s and the 1980s, 13 public sector electronic enterprises were created, including firms that are not primarily concerned with electronics but are major electronics producers, such as Bharat Heavy Electricals Limited (BHEL), Hindustan Machine Tools (HMT), and Hindustan Aeronautics Limited (HAL).[1] Nineteen state-level public sector electronics enterprises were set up; there was therefore one in most of the 25 Indian states. However, it is noteworthy that none of the three major electronics public enterprises—BEL, ITI, and ECIL—was or is under the direct control of the DOE, which controls only the software firm CMC, the chipmaker SCL, and the trading firm ET&T. The Department of Telecommunications (DOT) controls two manufacturing enterprises (in addition to two telecom service and one consultancy enterprises, which are not included in these 13), the Ministry of Industry controls four, the Ministry of Defense two, and the DAE and the Ministry of Science of Technology one each.

Aggregate Electronics Production and Growth Rates

Aggregate electronics production has grown from $231 million in 1971 to $993 million in 1981, to $5,257 million in 1990, and to $5,004 million in 1993. Tables 5.1 and 5.2 give electronics production and segment shares for the period 1971–93. Electronics equipment production—that is, aggregate production minus the output of components and Export Processing Zones (EPZs)—has grown from $178 million (1971) to $762 million (1981) to $3,760 million (1994).[2] These dollar conversions overstate Indian production, because domestic prices are much higher than world prices.

Aggregate electronics production in US$ grew, in nominal terms, at a compound rate of 15.7% during 1971–81, picking up to 22.8% over 1981–89 and to 24.8% during 1983–89—the period after the liberalizing policy changes of 1981–82 and before the downturn of the early 1990s. Taking electronic equipment alone, the acceleration of growth is more marked, with rates of 15.7% during 1971–81 and 23.4% during 1981–89, and 25.4% during the post-policy-shift boom years of 1983–89. The higher growth rates of the 1980s were due to the faster growth rates of consumer electronics, mainly television (38.8% over 1983–88), computers (39.8% over 1984–89), and telecom equipment (29.9% over 1985–89). Tables 5.1 and 5.2 give a disaggregated picture of output and output shares segment-wise.

Looking at Indian electronics in comparative perspective, it can be seen that it lost ground even in the 1980s, despite faster growth and narrowing of the gap with Brazil in the second half of the decade. Tables 5.3 to 5.6 contain comparability-adjusted figures for production, market size, exports, and imports, segment-wise, for India, Korea, and Brazil, during 1985–90. Starting with a higher level of output than Korea and a little behind Brazil in 1971, India was overtaken to the extent that by 1990, using comparable figures, India's output was $4.592 billion, Brazil's $12.020 billion, and Korea's $22.938 billion. By 1990, India's domestic electronics market was $5 billion, compared to Korea's and Brazil's roughly $14 billion each. However, Indian and Brazilian output and market size are much less unequal if one considers non-EDP output and market size only. Indian and Korean production and market size are less unequal if one takes into account only equipment, for Korea is a larger producer, exporter, and market (due to export production of equipment) for components.

Shifts in Intersegmental Output Shares

The weight of different segments in output has shifted significantly since 1971 (Table 5.2). Equipment as a whole remained steady in its share of total output during 1971–94 at 75–79%, peaking at 82.4% by 1987. Components, which had remained more or less steady at 23.1% (1971) to 20.2% (1980), fell to 14.8% by 1987 before rising to 18–19% by 1991–94. EPZ production was 2–4% of aggregate output. It is noteworthy that the industrial electronic equipment segment's share was an already high 46–48% during 1971–81, while the share of consumer electronics actually declined from 30.4% to 28.7% and of components from 23.1% to 20.2% during 1971–81, in contrast to the experience of Korea. Consumer electronics sharply increased its share of overall output in the 1980s, jumping from 29% in 1981 to 39% by 1987, reflecting the color-TV boom, before declining to 28% by 1994. Its share of equipment output moved from 40% in

Table 5.1: Indian Electronics Production by Segment (nearest US $ million)

Segment	*1971*	*1972*	*1973*	*1974*	*1975*	*1976*	*1977*	*1978*	*1979*	*1980*	*1981*	*1982*
Consumer	70	82	83	96	98	113	148	192	220	272	286	359
Telecom	53	62	75	86	119	125	146	156	158	235	178	269
Aero&Def	37	39	43	58	59	56	63	76	74	86	80	115
CIIE	17	22	28	42	70	71	117	145	161	178	184	201
Computer										26	34	55
EqptTotal	178	205	229	283	345	365	474	569	614	797	762	999
Component	53	58	66	89	90	89	106	143	167	207	200	226
EPZs					1	3	5	9	14	21	29	51
ElecTotal	231	263	294	372	435	458	582	721	796	1,025	993	1,277

Segment	*1983*	*1984*	*1985*	*1986*	*1987*	*1988*	*1989*	*1990*	*1991*	*1992*	*1993*	*1994*
Consumer	335	526	857	1,040	1,416	1,725	1,726	1,679	1,319	1,273	1,263	1,402
Telecom	267	284	308	400	548	647	878	931	851	972	1,033	1,116
Aero&Def	125	131	159	177	231	280	310	326	228	155	117	191
CIIE	249	295	327	416	529	673	746	800	620	617	558	637
Computer	77	81	125	222	289	349	432	468	365	374	344	414
EqptTotal	1,053	1,317	1,776	2,255	3,013	3,674	4,092	4,204	3,383	3,391	3,315	3,760
Component	228	267	331	404	540	735	888	868	770	839	807	940
EPZs	74	91	81	114	100	118	144	184	124	177	207	303
ElecTotal	1,356	1,674	2,189	2,774	3,653	4,527	5,123	5,257	4,276	4,408	4,329	5,004

Source: *Electronics Information and Planning*, Vol. 15, No. 9 (June 1988); Vol. 18, No. 3 (November 1991); Vol. 22, No. 10 (July 1995); *DOE Annual Report,* 1991–92.

Note: (1) Telecom includes Broadcasting Eqpt; Aero&Def = Aerospace & Defense; CIIE = Control Instrumentation and Industrial Electronics (included Computers—including software—until 1979); EqptTotal = Total Electronic Equipment; EPZ = Export Processing Zone (this does not represent all exports); Computer includes software unless production is in EPZ; ElecTotal = Total Electronic Production. (2) Exchange rates used for conversion are calendar-year-period averages from International Monetary Fund, *International Financial Statistics Yearbook* 1995, pp. 436–437.

1971 to 37% in 1981, rising to 48% in 1987 before declining to 37% in 1994. The share of communications declined from 22–23% during 1971–81 to 14.3% (1985) before rising to 24% by 1993. Computers, first counted as a separate category in 1980, rose in share from 3% in 1981 to 8% by the early 1990s.[3] In the 1980s, two major categories of industrial equipment involving technologically complex manufacture, CIIE and aerospace/defense, showed a decline in share. The relatively declining segments such as aerospace/defense, CIIE, and telecom equipment tended to be public enterprise-dominated (until the late 1980s), unlike the overwhelmingly private sector consumer electronics and computer industries, which increased their share over most of the 1980s.

The growth pattern of the post-1981 liberalization period explains this pattern of segmental growth rates and output shares. Following the liberalization of entry into TV production and the introduction of color broadcasting in 1982 and the expansion of TV broadcasting via satellite in 1983, a TV boom took place. A computer boom, mainly of PCs, followed the liberalization of entry and parts imports in 1984. By 1988–89, TV production actually declined, whereas import-substituting TV picture-tube production increased, raising the share of components. At the same time, the production of telecom switching equipment took off.

Imports and Market Size

Import statistics in India are hugely underestimated because of the prevalence of smuggling. Imports are estimated to comprise about 25–30% of final demand. They consist mainly of capital goods, components, parts, and electronic raw materials. Official figures, for what they are worth, indicate the relative size of imports to production in each category and can be used to estimate market size (production minus exports plus imports). According to comparability-adjusted figures, Indian electronics imports declined from 20% to 15% of output during 1985–90 and 19% to 14% of estimated market size over 1986–90 (Tables 5.7).[4] Disaggregating by major segments, as shown in the Table 5.7, import/production and import/market ratios have been lowest but rising during 1985–90 in consumer electronics, low and declining in telecom, declining in EDP production but steady in consumption, and highest but declining dramatically in components. Except for consumer electronics, import/production and import/market ratios rose again by 1993. However, it must be added that import intensity can vary quite dramatically in segments such as telecom, where imports consist substantially of capital equipment. It is also important to distinguish import/production ratios for each segment from import intensities of production, since the former includes only import of goods falling in that category, including final goods, but excludes component imputs, since components are a separate segment. Import intensity of production has remained fairly high, in line with global patterns at an average of 31% in the early 1990s: 64% in computers, 40% in software, 38% in components, and 16% in consumer electronics, which has experienced the most

Table 5.2: Change in Composition of Indian Electronics Production

Segment	*1971*	*1976*	*1981*	*1985*	*1986*	*1987*	*1988*	*1989*	*1990*	*1991*	*1992*	*1993*	*1994*
Consumer	30.4	24.7	28.7	38.7	36.8	38.6	38.1	34.3	31.9	30.9	28.9	29.2	28.0
Telecom	22.8	27.3	18.0	14.3	14.6	15.1	14.3	17.2	17.7	19.9	22.1	23.9	22.3
CIIE	7.5	15.6	18.7	15.2	15.2	14.5	14.8	14.6	15.2	14.5	14.0	12.9	12.7
Computer			3.3	5.8	8.1	7.9	7.7	8.4	8.9	8.5	8.5	7.9	8.3
Aero&Def	16.2	12.2	8.1	7.4	6.4	6.3	6.2	6.0	6.2	5.3	3.5	2.7	3.8
EqptTotal	76.9	79.8	76.8	81.4	81.1	82.4	81.1	80.5	80.0	79.1	77.0	76.6	75.2
Components	23.2	19.5	20.2	15.4	14.7	14.8	16.3	17.3	16.6	18.0	19.0	18.6	18.8
EPZs		0.7	3.0	3.2	4.2	2.8	2.6	2.8	3.5	2.9	4.0	4.8	6.1
Total	100	100	100	100	100	100	100	100	100	100	100	100	100

Source: *Electronics Information and Planning*, Vol. 15, No. 9 (June 1988), Vol. 18, No. 3 (November 1991), and Vol. 22, No. 10 (July 1995).

Note: Telecom includes Broadcasting Eqpt; Aero&Def = Aerospace & Defense; CIIE = Control Instrumentation and Industrial Electronics (this included Computers—including software—until 1979); EqptTotal = Total Electronic Equipment; EPZ = Export Processing Zone (this does not represent all exports); Computer includes software, unless production is in EPZ.

Table 5.3: Electronics Production in Brazil, India, and Korea (US$ million)

Segment	*Country*	*1985*	*1986*	*1987*	*1988*	*1989*	*1990*	*1991*	*1992*	*1993*
EDP	B	2,725	3,622	3,856	4,200	4,631	4,634	4,843	4,000	4,600
	I	133	249	313	414	523	585	455	453	428
	K	579	880	1,459	2,431	3,180	3,073	3,499	3,647	4,212
OffEqpt	B	241	266	284	310	319	315	318	302	312
	I	37	45	69	92	98	111	90	89	84
	K	79	105	196	207	276	265	253	256	290
ContInst	B	600	705	680	658	663	600	620	550	650
	I	170	222	232	300	327	311	261	232	215
	K	106	133	134	145	150	197	202	211	232
MedInd	B	234	234	242	242	229	210	225	195	225
	I	45	61	64	66	68	69	59	65	72
	K	47	52	64	97	145	171	186	192	243
Com&Mil	B	467	583	608	653	682	680	740	700	850
	I	256	286	457	503	592	658	519	501	479
	K	134	228	371	491	504	697	766	883	1,135
Telecom	B	743	945	1,200	1,420	1,520	1,320	1,280	1,100	1,250
	I	256	285	361	418	588	593	544	610	574
	K	769	876	1,183	1,245	1,890	1,756	1,884	1,798	1,831

TotIndElec	B	5,010	6,355	6,870	7,483	8,044	7,759	8,026	6,847	7,887
	I	894	1,149	1,497	1,794	2,197	2,327	1,928	1,950	1,852
	K	1,714	2,274	3,407	4,616	6,145	6,159	6,790	6,987	7,943
Consumer	B	1,353	1,746	2,056	2,085	2,296	2,151	2,270	2,000	2,303
	I	818	968	1,357	1,661	1,508	1,423	1,115	1,098	1,030
	K	2,086	3,136	4,726	6,138	6,576	6,312	6,697	6,363	6,689
Comp	B	1,910	2,345	2,401	2,548	2,464	2,110	2,433	2,236	2,303
	I	353	438	571	694	839	842	735	782	751
	K	2,640	3,860	5,550	7,892	9,807	10,467	11,959	12,793	14,530
TotalProd	B	8,273	10,446	11,327	12,116	12,804	12,020	12,729	11083	12,791
	I	2,065	2,555	3,425	4,149	4,544	4,592	3,778	3830	3,633
	K	6,440	9,270	13,683	18,646	22,528	22938	25,446	26143	29,162

Source: Elsevier Advanced Technology, Oxford, UK.

Note: B = Brazil; I = India; K = Korea; EDP = Electronic Data Processing, including software; OffEqpt = Office Equipment, i..e., typewriters, photocopiers, etc.; ContInst = Control and Instrumentation; MedInd = Medical and Industrial Electronic Equipment; Com&Mil = Communication and Military; Telecom = Telecom Equipment not in Com&Mil; TotalIndElec = Total Production excluding Consumer Electronics and Components; Comp = Components; TotProd = Total Production.

(1) These figures do not necessarily tally with national source figures for Brazil, India, Korea, but they agree with broad trends revealed by national figures. The figures are specially adjusted for comparability in terms of coverage, definitions, exchange-rate adjustments, shipment vs. production + inventory and other sources of error. (2) Internal consistency checks of Production + Imports = Exports + Apparent Domestic Market (i.e., Supply = Demand) disaggregated basis for all countries and years showed relatively small disrepancies mostly within a narrow band of error. India's export total for 1987 includes EPZs, but this is not reflected in the breakdown for that year.

Table 5.4: Electronics Markets of Brazil, India, and Korea (US$ million)

Segment	*Country*	*1985*	*1986*	*1987*	*1988*	*1989*	*1990*	*1991*	*1992*	*1993*
EDP	B	2,750	3,600	3,900	4,199	4,800	5,000	5,050	4,386	5,150
	I	160	297	402	501	571	615	519	530	545
	K	436	654	989	1,334	2,005	2,215	2,475	2,247	2,617
OffEqpt	B	243	263	278	297	317	328	342	324	336
	I	45	61	82	101	102	114	100	91	83
	K	77	102	99	129	197	209	219	199	221
ContInst	B	660	780	775	800	865	909	927	872	1,000
	I	243	254	294	388	420	409	359	344	313
	K	280	363	613	914	1,146	1,288	1,484	1,418	1,708
MedInd	B	255	284	272	281	326	375	353	314	358
	I	52	63	68	107	110	123	131	131	125
	K	109	119	128	164	208	292	346	317	443
Com&Mil	B	450	560	600	670	750	800	866	870	1,060
	I	289	354	492	525	612	676	540	520	523
	K	130	186	243	346	294	562	715	803	840
Telecom	B	730	910	1,180	1,413	1,520	1,324	1,298	1,168	1,330
	I	290	327	410	476	638	637	578	639	615
	K	775	727	797	975	1,539	1,454	1,507	1,436	1,267

TotIndElec	B	5,088	6,397	7,005	7,660	8,578	8,736	8,836	7,934	9,234
	I	1,078	1,356	1,747	2,097	2,452	2,573	2,228	2,256	2,204
	K	1,807	2,151	2,869	3,862	5,389	6,020	6,746	6,420	7,096
Consumer	B	1,332	1,730	1,802	1,852	2,051	2,031	2,114	1,816	2,098
	I	827	980	1,368	1,666	1,510	1,389	1,074	1,047	979
	K	726	864	1,160	1,470	2,045	1,962	2,281	2,224	2,481
Comp	B	2,022	2,522	2,604	2,776	3,037	3,022	3,140	2,739	3,179
	I	589	666	833	1,053	1,042	1,045	929	976	936
	K	1,706	2,873	4,303	5,388	6,322	6,042	6,782	6,397	8,321
TotMkt	B	8,442	10,649	11,411	12,288	13,666	13,789	14,090	12,489	14,511
	I	2,494	3,002	3,948	4,816	5,004	5,007	4,231	4,279	4,119
	K	4,239	5,888	8,332	10,670	13,756	14,024	15,809	15,041	17,898

Source: Elsevier Advanced Technology, Oxford, UK.

Note: B = Brazil; I = India; K = Korea; EDP = Electronic Data Processing, including software; OffEqpt = Office Equipment, i..e., typewriters, photocopiers, etc.; ContInst = Control and Instrumentation; MedInd = Medical and Industrial Electronic Equipment; Com&Mil = Communication and Military; Telecom = Telecommunication Equipment not in Com&Mil; TotalIndElec = Total Production excluding Consumer Electronics and Components; Comp = Components; TotMkt = Total Market.

(1) These figures do not necessarily tally with national source figures for Brazil, India, Korea, but they agree with broad trends revealed by national figures. The figures are specially adjusted for comparability in terms of coverage, definitions, exchange-rate adjustments, shipment vs. production + inventory and other sources of error. (2) Internal consistency checks of Production + Imports = Exports + Apparent Domestic Market (i.e., Supply = Demand) disaggregated basis for all countries and years showed relatively small disrepancies mostly within a narrow band of error. India's export total for 1987 includes EPZs, but this is not reflected in the breakdown for that year.

Table 5.5: Comparative Exports of Brazil, India, and Korea (US$ million)

Segment	*Country*	*1985*	*1986*	*1987*	*1988*	*1989*	*1990*	*1991*	*1992*	*1993*
EDP	B	189	180	137	217	175	93	200	196	250
	I	3	16	4	55	150	200	123	190	74
	K	535	920	1,466	2,353	2,506	2,447	2,648	2,808	3,157
OffEqpt	B	16	23	24	22	27	27	37	10	12
	I	1	1	10	14	12	11	9	11	10
	K	27	38	121	110	122	113	104	108	125
ContInst	B	30	40	23	23	35	42	60	63	70
	I	3	4	3	14	9	9	7	7	7
	K	41	49	81	127	179	210	158	229	177
MedInd	B	4	4	5	4	4	3	8	9	12
	I	1	6	3	65	32	37	35	9	4
	K	11	14	18	48	78	49	59	64	21
Com&Mil	B	63	78	48	38	6	6	40	7	20
	I	2	2	0	11	6	7	8	5	6
	K	81	122	217	315	341	409	419	442	623

Telecom	B	23	45	51	43	41	71	90	35	40
	I	1	2	1	4	0	1	1	2	2
	K	176	270	458	569	582	539	564	542	753
Consumer	B	71	80	297	297	394	347	383	318	352
	I	3	3	9	51	27	33	35	53	55
	K	1,507	2,480	3,930	4,744	4,511	4,491	4,535	4,318	4,392
Compt	B	150	198	149	170	146	136	182	220	327
	I	13	8	33	74	57	72	53	70	68
	K	1,560	2,479	3,536	5,494	6,583	7,424	8,924	9,922	10,398
TotMkt	B	546	648	734	814	828	725	1,000	858	1,083
	I	29	40	164	289	293	369	271	247	225
	K	3,938	6,372	9,827	13,760	14,902	15,682	17,391	18,433	19,646

Source: Elsevier Advanced Technology, Oxford, UK.

Note: B = Brazil; I = India; K = Korea; EDP = Electronic Data Processing, including software; OffEqpt = Office Equipment, i..e., typewriters, photocopiers, etc.; ContInst = Control and Instrumentation; MedInd = Medical and Industrial Elec Equipment; Com & Mil = Communication and Military; Telecom = Telecommunication Equipment not in Com&Mil; TotalIndElec = Total Production excluding Consumer Electronics and Components; Comp = Components; TotMkt = Total Market.

(1) These figures do not necessarily tally with national source figures for Brazil, India, Korea, but they agree with broad trends revealed by national figures. The figures are specially adjusted for comparability in terms of coverage, definitions, exchange-rate adjustments, shipment vs. production + inventory and other sources of error. (2) Internal consistency checks of Production + Imports = Exports + Apparent Domestic Market (i.e., Supply = Demand) disaggregated basis for all countries and years showed relatively small disrepancies mostly within a narrow band of error. India's export total for 1987 includes EPZs, but this is not reflected in the breakdown for that year.

Table 5.6: Comparative Imports of Brazil, India, and Korea (US$ million)

Segment	*Country*	*1985*	*1986*	*1987*	*1988*	*1989*	*1990*	*1991*	*1992*	*1993*
EDP	B	174	158	181	216	344	459	600	582	800
	I	73	71	142	209	210	181	141	168	191
	K	392	694	996	1,256	1,331	1,489	1,624	1,408	1,562
OffEqpt	B	29	32	18	9	25	40	71	32	36
	I	8	9	19	24	15	11	10	13	9
	K	25	32	24	32	43	57	70	51	56
ContInst	B	90	115	118	165	237	351	400	385	420
	I	27	48	73	193	70	97	93	119	104
	K	215	279	560	896	1,175	1,301	1,440	1,436	1,653
MedInd	B	47	54	35	43	101	168	210	128	145
	I	15	13	13	69	55	51	44	76	57
	K	73	81	82	115	141	170	219	189	221
Com&Mil	B	46	55	40	55	74	126	170	177	230
	I	15	48	31	150	25	23	17	23	49
	K	77	80	89	170	131	274	323	362	328

Telecom	B	10	10	31	36	41	67	100	103	120
	I	17	47	50	107	15	14	8	31	43
	K	182	121	72	181	231	237	187	180	189
Consumer	B	50	64	43	64	149	248	323	134	147
	I	8	21	69	41	44	67	831	3	4
	K	130	199	300	77	141	147	165	179	184
Comp	B	263	369	352	396	733	1,090	1,251	723	905
	I	247	322	359	397	176	234	209	264	254
	K	819	1,494	2,289	2,940	1,218	2,999	3,664	3,526	4,189
TotMkt	B	709	857	818	984	1,704	2,549	3,123	2,264	2,803
	I	410	578	758	1,189	610	679	553	696	711
	K	1,913	2,980	4,412	5,667	6,291	6,674	7,692	7,331	8,382

Source: Elsevier Advanced Technology, Oxford, UK.

Note: B = Brazil; I = India; K = Korea; EDP = Electronic Data Processing, including software; OffEqpt = Office Equipment, i..e., typewriters, photocopiers, etc.; ContInst = Control and Instrumentation; MedInd = Medical and Industrial Elec Equipment; Com & Mil = Communication and Military; Telecom = Telecommunication Equipment not in Com&Mil; TotalIndElec = Total Production excluding Consumer Electronics and Components; Comp = Components; TotMkt = Total Market.

(1) These figures do not necessarily tally with national source figures for Brazil, India, Korea, but they agree with broad trends revealed by national figures. The figures are specially adjusted for comparability in terms of coverage, definitions, exchange-rate adjustments, shipment vs. production + inventory and other sources of error. (2) Internal consistency checks of Production + Imports = Exports + Apparent Domestic Market (i.e., Supply = Demand) disaggregated basis for all countries and years showed relatively small disrepancies mostly within a narrow band of error. India's export total for 1987 includes EPZs, but this is not reflected in the breakdown for that year.

Table 5.7: Import/Production and Import/Market Ratios for Indian Electronics, 1985-1993

	1985		*1986*		*1990*		*1993*	
Segment	*Impt/ Prod*	*Impt/ Mkt*	*Impt/ Prod*	*Impt/ Mkt*	*Impt/ Prod*	*Impt/ Mkt*	*Impt/ Prod*	*Impt/ Mkt*
EDP	54.9	45.6	28.5	23.9	30.9	29.4	44.6	35.0
Telecom	6.6	5.9	16.5	14.4	2.4	2.2	7.5	7.0
TotIndElec	17.3	14.4	20.4	17.3	16.2	15.5	24.5	20.6
Consumer	1.0	1.0	2.2	2.1	4.7	4.8	0.4	0.4
Component	70.0	41.9	73.5	48.3	27.8	22.4	33.8	27.1
Total	19.9	16.4	22.6	19.3	14.8	13.6	19.6	17.3

Source: Computed from Elsevier Advanced Technology, Oxford, UK.

Notes: Impt = Imports; Prod = Production; Mkt = Market; EDP = Electronic Data Processing (computers including software); TotIndElec = Total Industrial Electronics

ISI.[5] At $5 billion in 1990, India's domestic market size was just over a third of Brazil's and Korea's, which were $14 billion each. This reflects the fact that electronics demand is highly sensitive to per capita income and India's vast population does not automatically imply a large domestic market, especially with higher-than-world prices, although—and this is important for policy—it can potentially be.[6]

Exports

Exports rose from $29 million in 1985 to $369 million in 1990 to $532 million in 1993—minuscule compared to Korea's $15,682 million in 1990 and $19,646 million in 1993, and to Brazil's $725 million in 1990 and $1,083 million in 1993. These figures indicate an impressive rise, although export accounted for only 8–9% of output by 1990–91. Exports can be divided into those from the EPZs and those from the Domestic Tariff Area (DTA). EPZ exports grew faster than DTA exports and overtook the latter in 1982, remaining ahead until 1987. About a third of exports during 1986–91 came from EPZs. Computer software and components are the largest single export items from both the DTA and the EPZs. Software has tended to dominate DTA, and components EPZ exports. The share of computer software in exports—mainly to the United States—has risen over 1987–93 from about a fifth to close to half. Electronics hardware exports, largely to the former Soviet Union, have tended to decline because of the latter's collapse. The overall picture, however, indicates extreme uncompetitiveness.

The Six Major Segments of the Indian Electronics Industry

A detailed look at the six major subdivisions of the Indian electronics industry is instructive. Table 5.8 gives the output volumes of major products during 1981–93. Consumer electronics in India is dominated by the TV industry, which accounted for as much as 70% of output in 1987, declining to a little over half in 1993.[7] The spurt of the 1980s has been mainly due to b&w television in the first half of the decade and to color television since then, the latter overtaking the former in value in 1986. This development was made possible by the rapid growth of the population covered by TV broadcasting since 1982 (Table 5.9). The number of b&w TV sets produced increased from 0.66 million in 1983, to 3.2 million in 1987, to 4.1 million by 1993; and color TV sets from 50,000 in 1983 to 1.10 million in 1987, followed by a recession, from which 1987's peak production was surpassed only in 1994. The production of radio receivers increased from 6 million in 1983 to 8 million in 1987 but declined to 5 million in 1993, following international trends with a 5–10 year lag, but offset by increased production of radio-cassette recorders of 6.5 million in 1993. Electronic watch production rose from 28,000 in 1983, to 1.30 million in 1987, to 6.7 million in 1993.

The industrial electronic equipment segments taken together have bulked largest in Indian electronics, accounting for 47% of total output and 61% of equipment output in 1971, 51% and 66% in 1980, and 48% and 62% in 1992 (Tables 5.1 and 5.2). The industrial equipment segments consist of communications and broadcasting equipment, CIIE, computers, and aerospace/defense, and they have been the principal focus of official promotional efforts. The three largest and oldest public enterprises in electronics—ITI, BEL, and ECIL—are the leading firms in these segments. Telecom (until the early 1990s) and strategic electronics have been public sector-dominated. The rising share of industrial electronic equipment is revealed even more clearly if we look at civilian industrial electronic equipment alone, omitting stagnant aerospace/defense. Its share rose from 39% of equipment output in 1971, to 48% in 1981, to 58% by 1992.

The communications and broadcasting (henceforth telecom) segment is the oldest part of the industrial equipment segment. Telecom has been a public sector preserve by policy, with ITI the monopoly producer of equipment for the state-owned telephone network—except for subscriber-end equipment since 1984—until the liberalization of the 1990s. A broad range of telecom equipment for switching, transmission, and subscriber premises is manufactured, including, since 1986, digital electronic switching systems (ESS) with a range of foreign technology suppliers.

After years of stagnation, ITI has undergone rapid growth and modernization since the signing of the deal, in 1982, for transfer of technology to manufacture digital exchanges with CIT-Alcatel. Full digitalization of the telecom network is the long-term objective. Originally, the plan was to phase in the manufacture of large digital exchanges designed indigenously by the Centre for the Development of Telematics (CDOT). However, since 1990, following a change of government,

Table 5.8: Production of Major Electronic Products in Quantity, 1981–1993

Item	*Unit*	*1981*	*1982*	*1983*	*1984*	*1985*	*1986*	*1987*	*1988*	*1989*	*1990*	*1991*	*1992*	*1993*
Radios	MNo	6.58	7.09	5.99	6.23	7.46	7.50	7.99	8.00	8.00	6.00	6.0	5.5	5.0
Tape Rec	MNo	0.38	0.58	0.80	1.12	1.69	1.87	2.74	2.97	3.34	2.12	1.74	0.19	0.08
Tape Rec Radio	MNo	0.13	0.14	0.09	0.10	0.30	0.64	0.32	0.52	0.66	2.91	3.60	5.8	6.36
B&W TV	MNo	0.44	0.57	0.66	1.00	1.80	2.15	3.20	4.40	4.00	3.60	3.10	3.40	4.10
Color TV	MNo		0.07	0.05	0.28	0.68	0.85	1.10	1.30	1.20	1.20	0.88	0.83	1.07
Mini/Micro-Computers	1000No	0.486	0.520	1.717	3.060	7.516	29.61	45.87	60.35	41.99	95.64	99.68	90.0	111.0
Large Elec Exchanges	MLine	0.02	0.02	0.01	0.03	0.03	0.14	0.23	0.27	0.62	0.68	0.99	1.17	1.55
Electro-MechPhone	MNo	0.33	0.50	0.59	0.70	0.66	0.75	0.76	0.91	0.73	0.70	0.59	0.85	
Elec Phone	MNo						0.05	0.05	0.35	0.80	1.31	0.74	1.17	2.50
B&W TV Tube	MNo	0.26	0.36	0.48	0.81	1.45	1.91	3.19	4.85	4.57	3.85	3.45	3.75	4.59
CTV Tube	MNo							0.08	0.38	1.08	1.23	1.12	0.90	1.03
IC Linear	MNo	0.09	0.07	0.10	0.27	0.37	1.23	1.39	2.89	6.55	13.02	10.41	3.02	0.12
IC LSI/VLSI	MNo				0.53	1.66	1.87	5.24	6.70	1.13	2.69	3.18	10.91	6.99
IC Hybrid	MNo	0.03	0.05	0.06	0.07	0.35	1.00	2.14	2.97	5.72	6.95	5.42	13.82	35.87

Source: Electronics Information and Planning, Vol. 19, No. 2 (November 1991); Vol. 20, No. 8 (May 1993); Vol. 21, No. 9 (June 1994).

Note: 1000No = thousands; MNo = millions; MLine = million lines; IC = integrated circuit; rec = Recorder; Elec = Electronic; ElectroMech = Electromechanical.

Table 5.9: TV Production in India and Transmission Coverage, 1970–1990

			TV production (millions)		
		Population	*Black-and-White TV*		
Year	*TV Centers*	*Coverage (%)*	*51-cm size*	*36-cm size*	*Color TV*
1970	1		0.014		
1971	1		0.016		
1972	2		0.030		
1973	5		0.075		
1974	5		0.075		
1975	10		0.097		
1976	10		0.144		
1977	14		0.238		
1978	16		0.270		
1979	18		0.311		
1980	18	25	0.370		
1981	18	25	0.440		0.010
1982	19	25.69	0.570		0.070
1983	41	26.35	0.660		0.050
1984	46	30.37	0.825	0.175	0.280
1985	175	56.18	1.360	0.440	0.680
1986	185	66.68	1.330	0.820	0.850
1987	224	70.30	1.375	1.625	1.100
1988	274	72.00	1.700	2.700	1.300
1989	508	73.90	1.300	2.700	1.200
1990	520	76.00	3.500		1.200

Source: Electronics Information and Planning, Vol. 18, No 11 (August 1991), Table 4, p. 608.

large-scale imports have been resorted to. With the increase in production of electronic exchanges since the late 1980s, the bottleneck on expansion of telecom has started to be overcome. Telephone connections had grown at only 8.4% annually during 1979–89, to only 4.6 million lines, expanding faster to over 8.63 million by late 1994—still one of the world's lowest densities.[8] Since the late 1980s, the expansion of specialized datacom networks has picked up, representing a shift in the state's role to infrastructure provider. Telecom policy has been in a state of transition since the beginning of the 1990s, with the drift being toward breaking up the DOT monopoly, plugging the domestic production shortfall with imports, separation of policy, regulation, and service provision, and freer private and foreign entry; and, by 1994, telecom equipment manufacture was thrown open to MNCs, as were newly introduced cellular and, later, basic service in joint venture form.

The Indian computer industry has a checkered history. Originally dominated by MNCs such as IBM and ICL, it was progressively indigenized, first by ECIL's TDC series of minis in the 1970s, then, in the 1980s, by domestic private firms, especially after the New Computer Policy of November 1984. In the early 1970s, ECIL designed a minicomputer—the TDC–12—which, with its successor, the TDC–16, was on the market by 1972–73. Their successors, the TDC–312 and 316, were on the market by 1974–75. The Minicomputer Panel Report of September 1973 saw an opportunity for domestic computer manufacturers based on the systems engineering route (putting together imported chips and peripherals according to one's own design). The boom really took place after the 1984 policy, volumes rising especially for the PC range as prices fell. The number of PCs produced rose from 7,500 in 1985 to 110,000 by 1993. However, this is minuscule by developed-world or even Korean or Brazilian standards. And Indian prices, despite falling steeply, remained well above world prices.

This rapid growth has come from the explosive growth of demand for PCs. The number of manufacturers—initially mainly assemblers of imported parts—mushroomed in the first half of the decade and is now getting concentrated. Until the liberalization of 1991, there was one manufacturer of true mainframes—ECIL, making Control Data's Cyber 180/810 and 830 under license—and two of superminis—ECIL and ICIM (formerly ICL India) and ICIM (ICL India). By the early 1990s, new products were being introduced with a lag of only months.

Computer software has been a growth area, especially since the liberal New Software policy of December 1986. Software exports, dominated by two firms, Tata Consultancy Services (TCS) and Tata Unisys (earlier Tata Burroughs), have risen to $226 million in 1993. The former firm operates out of the DTA and the latter out of the Santa Cruz Electronics Export Processing Zone (SEEPZ). India has become one of the leading developing-country software exporters. It is the only segment of Indian electronics that is predominantly world-market-oriented. Exports are dominated by on-site labor-intensive coding and systems analysis and design jobs.

The control instrumentation and industrial electronics (CIIE) category is extremely heterogeneous. It consists of test and measuring instruments, analytical instruments, special application instruments, medical electronics, process-control equipment, and other miscellaneous industrial electronic equipment. The demand for this segment comes from large industrial, electric power, and transportation enterprises, which are mostly in the public sector. ECIL and BEL are major producers of CIIE, the former specializing in nuclear instrumentation, the latter in control systems for power plants. Power electronics equipment has been the single largest chunk of CIIE.

Aerospace and defense electronics has always been administratively separate from the rest of the electronics industry in that it comes under the jurisdiction of the Ministry of Defense, not under the DOE. This segment has grown relatively slowly, its growth in the 1970s being actually negative in real terms. Production is reserved for the public sector, led by BEL. Defense communication equipment and radars are the major products.

Electronic components have also been a focus of promotional thrusts, mainly of an ISI type. However, output remains tiny at $780 million in 1992 by world—even other NIC—standards. This figure overstates its size, as Indian prices are over twice the world price for most components. The components industry is very diversified, with each product being produced in tiny volumes by world standards. The largest single segments in recent years have been electron tubes (mainly b&w TV picture tubes, now being overtaken by color picture tubes) and passive components. However, if one considers components specific to consumer electronics other than ICs, PCBs, and passive components, then over one-third to one-half of output consists of these. India is about 75% self-sufficient in consumer-grade components (especially b&w picture tubes) but relies on imports for two-thirds of its professional-grade components, including 95% of its ICs, and most electronics raw materials are imported. All semiconductors represented only an eighth of component output in 1993. However, even here, despite tiny demand, ISI has been the strategy.

The semiconductor, and especially the IC industry, has been sought to be vigorously promoted. Small- and medium-scale ICs using bipolar technology have been produced at BEL since the 1970s on a full-process (including diffusion step) basis. The IC effort took a quantum jump with the coming on stream of the SCL, the semiconductor "national champion," at Chandigarh in 1984. There was a vigorous push toward full-process IC design and fabrication in SCL, based initially on imported CMOS technology, but with a focus on acquiring ASIC design and manufacturing capability for applications in defense, space, and telecom. Though ASICs have lower technological and financial entry barriers, even for these the levels of domestic demand and investment were insufficient and export tie-ups on OEM or joint venture basis nonexistent. SCL was destroyed in a fire in 1989 and is being rebuilt mainly for strategic self-reliance, although reconstruction has been on hold since November 1994. It is now planned to be partially privatized, offering its manufacturing facility to Texas Instruments, and will try to become an ASIC design center.

Semiconductor output reached $97 million in 1992, of which ICs (linear and digital, bipolar, and MOS) accounted for $60 million. The numbers produced were extremely tiny—for example, 11 million LSI/VLSI devices, a tiny fraction of the minimum economic scale for any one device in ICs.[9] Discrete devices such as diodes and transistors are produced by several mostly private firms, but including BEL and ECIL.

THE PUBLIC SECTOR, FOREIGN FIRMS, AND THE DOMESTIC PRIVATE SECTOR, LARGE- AND SMALL-SCALE, IN INDIAN ELECTRONICS

The Public Sector

The Indian electronics industry has been spearheaded in its development, especially in the more complex segments, by public enterprise. The public sector's focus has been on industrial equipment and components. ECIL is the most diversified, being active in all segments. BEL is active in all but computers and CIIE. HAL specializes in defense/aerospace and telecom. SCL (LSI/VLSI ICs) and Central Electronics Limited (solar photovoltaic cells) specialize in components. ITI, HCL, and HTL are confined to telecom, IL and BHEL to CIIE. HMT (digital watches), BEL (TV glass shells), and ECIL (color television) have made forays into consumer products as a sideline. There are also state-government-owned enterprises, some of which are significant producers.

However, even the largest of these are small by world standards and highly diversified. ITI, BEL, and ECIL have sales of only on the order of hundreds of millions of dollars. By contrast, the sales of any large Western, Japanese, or Korean firm was of the order of billions of dollars by 1985 and tens of billions by the mid-1990s.[10]

The public sector's share of gross electronics output fell from 56% (1971) to 43% (1981) and then to 30% (1988–89), reflecting a relative de-emphasis in the 1980s, with the private sector-dominated computer and consumer electronics industries exhibiting the fastest growth.[11] The state public sector corporations accounted for about 20% of total public sector output in 1988–89, up from 12% in 1981. However, the loss of share in the 1970s despite the public sector-led ISI thrust in the industrial equipment segments reflects the rise of the consumer electronics industry, which was reserved for the small-scale industry (SSI) sector. As of 1980, the public sector accounted for all of aerospace/defense and telecom and 34.5% of computers and CIIE; in industrial equipment as a whole, it had a 68% share. By 1987, it had a share of 7% in consumer electronics, 27% in CIIE, 12% in computers, 20% in components, 93% in telecom, and 100% in aerospace/defense.

The Private Sector

Foreign Firms

The private sector now dominates Indian electronics production, with 68% of output by 1987 and a much higher share by the early 1990s, following liberalization. The sector can be subdivided into foreign and domestic firms, and the latter, in turn, into large- and small-scale industry. In the Indian regulatory context, until 1991 "foreign" meant having 40% or more foreign-held equity, as per the rules of the Foreign Exchange Regulation Act (FERA) of 1973. FERA also allowed foreign firms defined as above to be licensed only if they brought in advanced technologies or were 100% export-oriented, as in the EPZs. They were strongly discouraged in the 1970s, especially in the profitable consumer electronics industry. FDI in the Indian electronics industry was virtually nil during the period concerned, as even Indian EPZs were comparatively unattractive.

IBM, which dominated the Indian computer industry, pulled out in 1978 over the FERA equity dilution issue. The only other major foreign firms in Indian electronics, Philips and ICL (UK), agreed to dilute foreign equity to below 40% and become Indian companies under the law, but they remained foreign-controlled. Siemens, which has foreign majority equity, is present in India, but not in electronics. Unisys (formerly Burroughs) has a 50/50 joint venture with Tata in SEEPZ to produce software for export. MNC shares in Indian electronics output dropped from 10% to 2% between 1972 and 1977 as a result primarily of FERA-induced disinvestment. Most subsequent collaborations involved only technology, not equity. Philips was restricted to components and, since 1984, consumer electronics, and ICL (now International Computers Indian Manufacture—ICIM) to computers, both being considered Indian firms after dilution of foreign equity to below 40%. Except for software exports (Texas Instruments, Citicorp), foreign investment in Indian electronics was negligible. However, since 1991 there have been large numbers of foreign entrants, mostly in joint ventures with the Indian private sector, including IBM, once again (in alliance with Tata in software), Hewlett Packard and Groupe Bull in computer hardware, and Alcatel and Fujitsu in digital switching, to name only a few.

Foreign technology import was restricted after 1970–71 by clauses that discouraged technology suppliers. However, the number of collaborations increased steadily, from 15 in 1978 to 90 in 1984, then jumping to 173 in 1985, when foreign collaboration rules were liberalized. The industrial equipment and components segments account for the overwhelming majority of collaborations, consumer electronics the least. Because Indian terms were unattractive, they also tended to be of low "quality"—that is, they did not involve as much transfer of technology or the latest in technology as, say, Korea or Brazil in their imports of technology.

Small-Scale Industry

The private sector in India is subdivided into the large- and medium-scale or "organized" private sector and the small-scale industry (henceforth SSI) sector.[12] SSI means all industry with fixed assets valued below a certain ceiling. This concept evolved in the 1960s, and since then the ceiling has been periodically revised upwards; in the early 1990s, it stood at Rs 3.5 million for independent SSI firms and Rs 4.5 million for ancillary (subcontracting) firms. A long list of products, especially in consumer electronics and components, was reserved for SSI in the 1970s. Though considerable dereservation has taken place, 11 items—mostly consumer-grade components—were still reserved until the post-1991 period. Because of this, the SSI sector increased its share of output sharply, from 20% to 29% between 1971 and 1981, and in consumer electronics from 36% to 49%. Its shares of computers and CIIE remained constant at 34–35%, and of industrial equipment as a whole they rose from 6% to 16% and of components from 23% to 34%, between 1971 and 1981. By 1987, the share of SSI in consumer electronics had actually risen to 58%, in CIIE to 31%, computers to 33%, and components to 36%, despite liberalization. However, by 1988–89, SSI share of overall output was 38%, or more than half the private sector's 70%.[13] These statistics, despite being suspect as many larger firms continue to be registered as SSI by underdeclaring assets and sales so as to gain various fiscal and financial benefits, are a rough indicator of the policy-induced presence of SSI.

Large Domestic Firms

The share of the "organized" (non-SSI) private sector, both Indian and foreign, in total output has been at 31–32%. In consumer electronics, the organized private sector had 42% (1981), falling to 35% (1987). In telecom, its share increased from 3% (1981) to 4% (1987). In CIIE it went from 30% (1981) to 33% (1987), and in components it rose from 38% to 43%. In computers, however, it went from 63% (1981) to 54% (1987)—a fall in share despite tremendous growth of this sector and considerable liberalization. The figure for computers is suspect and contradicts figures by *Dataquest* magazine, showing organized private sector firms accounting for at least 70–80% of industry hardware revenues. By 1987, the organized private sector dominated only computers and components.

The overall tendency since the late 1980s on has been for the shares of the public and SSI sectors to fall and for those of the private sector, including foreign firms, to rise with liberalization and growth led by consumer electronics, computers, software, and, by the early 1990s, telecom. By 1988–89, just before the 1991 liberalization, the public sector's share of total output was 30%, of equipment output 33%, components output 20%. It had a monopoly only in strategic electronics and was dominant—but no longer a monopoly—in only telecom at 84%. It had 25% of CIIE, 15% of computers (but only 6% of the dynamic software export segment), 10% of consumer electronics. The private sector share of total output had risen to 70%—38% in SSI and 32% in the organized private sector.[14]

INVESTMENT AND TECHNOLOGY DEVELOPMENT IN INDIAN ELECTRONICS

Investment

The investment pattern in Indian electronics has so far been one of fragmentation of investment into subcritically small units, due to restrictive licensing and SSI policies. This is so not only in SSI-reserved sectors such as consumer electronics, but also in public investment in manufacturing and R&D programs, even in such scale-sensitive segments as semiconductors. The result has been the spawning of a vast number of subcritically small firms, not just in SSI but also in the public sector.

The Seventh Five-Year Plan 1985–90, covering the boom period of the 1980s liberalization just before the slowdown and reforms of 1990 and 1991, saw total cumulative investment in electronics production of Rs 24,400 million, or well under $2 billion. Components accounted for 45% of this and communications for another 14%.[15] The public sector (DOE, other ministries and departments, public enterprises) had been expected to invest 60% and the private sector 40%. However, in communications the public sector had been expected to put up 81%, and in components 40% of the required investments. But the Seventh Plan outlay for telecom (not only in production) was scaled down from Rs 130 billion to Rs 47.1 billion—just over half the estimated required investment for meeting Plan targets for telecom service expansion.[16] Thus ITI lacked the necessary demand for equipment from the DOT. Component demand depends on the growth of the equipment segments, and telecom is especially important for ICs, given that the IC industry is to specialize in semicustom and custom chips geared to telecom equipment and not to standard memories and microprocessors for computers. It appears now that the Plan's estimates were overoptimistic for telecom, computers, and CIIE, and an underestimate for consumer electronics.

In the 1985–90 Plan period, capacity minimums were laid down for scale economies for new projects.[17] However, most of these fell short of world scales and World Bank-suggested scales, particularly in ICs. In ICs, the minimum economic scale for full-cycle production was $200 million at plant level and $500 million at firm level in 1985. Compared to these figures, the total projected investment for all components for the Plan period was Rs 8,000 million—that is, well below $800 million. India's LSI/VLSI "national champion," SCL, had a cumulative total investment of under Rs 500 million in 1985–86 (under $40m) and produced under 12 million ICs of all types, including just over 5 million LSI/VLSI. In 1986–87 and 1987–88, the outlays approved for SCL were only Rs 150 million and Rs 100 million, clearly far short of the then necessary minimum of $200 million for wafer fabrication for LSI/VLSI.[18] Before SCL burnt down, India's total LSI/VLSI IC production was under 7 million devices of all types.

In general it would appear that despite a policy shift toward scale minimums, liberalized entry, and stiffer competitive pressures and despite a sharp increase in private sector investments, the problem of suboptimal, fragmented investment in both manufacturing and R&D was not surmounted. This criticism applies to the DOE's own annual plan outlays. For example, in 1987–88 the DOE's Plan outlay of only Rs 1309 million and its approved 1988–89 outlay of just Rs 1,200 million were to be spread over 36 manufacturing and R&D programs, with SCL (in VLSI ICs!) getting Rs 100 million and Rs 40 million, respectively, and lacking adequate internal resources to plow back; CMC getting Rs 70 million and

Rs 40 million, respectively; CDOT getting Rs 30 million and Rs 55 million, and even a Fifth Generation Supermini Computer Design Program (!) getting just Rs 25 million and Rs 40 million, respectively, with the rupee declining to Rs 13–14 to the dollar. This pattern had not changed by 1994–95.[19] Relatively few scale-sensitive areas can be said to be potentially competitive—perhaps b&w televisions and picture tubes. Scale-sensitive products are inappropriate for India, given its small domestic market and the already established presence in world markets of the giant Japanese and other East Asian NIC firms. Additionally, labor-cost advantages in electronics assembly have been rendered irrelevant by the growing trend toward automation and new technologies, such as surface mounting.

Scientific and Technological Efforts

There has been a broad-based effort in scientific and technological development in India, with a strong emphasis on self-reliance, with the goal of being able to generate technology indigenously rather than having to import successive rounds of technology, and also to save on scarce foreign exchange. Electronics fits the pattern. The bulk of R&D activity has taken place in government-funded institutions and public sector firms. Very little R&D takes place in the private sector, except for design efforts in some computer and software firms. Public sector electronics R&D concentrates on activities of a more basic research character for acquiring technological depth rather than on adaptive design and production technology for commercially viable products. Research institutions have very weak links with industry and lack manufacturing facilities.

The result has been a considerable accumulation of technological capability, including in-depth research activities, but few commercially viable products transferred to industry; certainly, no internationally competitive ones. India's restrictive policies on royalty payments, combined with the lack of domestic competition, have also resulted in relatively few and "poor-quality," bare-minimum technology transfer agreements. Import of technology is inescapable in this fast-changing industry. An overemphasis on indigenous technology development may well cause the obsolescence of the products of indigenous efforts by the time they materialize, and hence be responsible for a technological lag.

The major Indian research institutes are: the Central Electronics Engineering Research Institute (CEERI), Pilani; the Central Scientific Instruments Organization (CSIO), Chandigarh; the National Physical Laboratory (NPL), New Delhi; the Bhabha Atomic Research Center (BARC), Bombay; the National Aeronautical Laboratory (NAL), Bangalore; the Indian Institute of Science (IISc), Bangalore; the five Indian Institutes of Technology (IITs); the Electronics Research and Development Center (ERDC), Trivandrum; and the Electronics and Radar Development Establishment (LRDE), Bangalore. Between 1973 and

1987 (the exchange rate moving roughly from Rs 7 to Rs 13/$) the DOE spent Rs 635 million on 400 projects, spread over 100 R&D labs (over 100 projects), educational institutions (over 200 projects), and public sector firms (about 100 projects), of which 275 had then been completed.[20] The Indian electronics R&D base is spread across the Departments of Electronics, Atomic Energy, Space, Communications, Defense, and Industry and academia.

The total R&D investments in the Fifth (1975–79), Sixth (1980–85), and Seventh (1985–90) Plans were Rs 1,640 million, Rs 4,450 million, and Rs 4,234 million, respectively—the last figure being for public investment only, with funds from DOE's programs accounting for 45%. The rest presumably comes from Defense, Space, Atomic Energy, Telecommunications, and other departments. During 1985–90, the DOE set up the National Microelectronics Council (NMC), the Council for the Development of Materials (CDOME), the Center for the Development of Advanced Computing Technology (CDAC), and the Society for Applied Microwave Engineering and Electronics Research (SAMEER).[21]

A major indigenous technology initiative was launched with the setting-up of the Centre for Development of Telematics (CDOT) in 1984, headed by a US-returned expert, Satyen G. Pitroda, to develop a family of cheaper, indigenously designed digital exchanges by building upward from smaller modules, starting with a Rural Automatic Exchange (RAX) adapted to Indian conditions and an Electronic Private Automatic Branch Exchange (EPABX), and eventually a Main Automatic Exchange (MAX) to replace the French CIT-Alcatel E–10-B large digital exchanges made by ITI, under a 1982 deal, from 1986 on. Pitroda managed to keep the state out of production, as in the cases of CPqD and ETRI, and CDOT, technology was licensed out to private firms. However, again there was no standardization of equipment, as three foreign technologies were each permitted to be imported for EPABX and trunk exchanges production by other firms in the mid-1980s.

A major effort was launched in 1988 by the newly set up Centre for Development of Advanced Computing (CDAC) at Pune, technically successful by 1992, to design and build a parallel processing supercomputer (PARAM) based on the Inmos transputer chip capable of a speed of up to 1,000 megaflops. The PARAM has been exported to Canada and Germany. Two other parallel processing machines have been designed and built at the Defence Research and Development Organization (DRDO) and the National Aeronautical Laboratory (NAL).

Two tendencies have been evident. One is the fragmentation of funding into subcritically small amounts in a very large number of projects, many of a very ambitious and basic character, requiring vastly greater resources to bear fruit.

The second tendency has been the bias in favor of basic research—often very ambitious, given the level of development of the industry. The basic objective apart from defense-oriented projects was to acquire an across-the-board "total system capability" in the core infrastructural sectors of the economy. Such an objective dictated an R&D program that was both broad-based and in-depth, with the result that limited resources were thinly spread. As Ashok Parthasarathi, who was Secretary to the EC and Joint Secretary and Additional Secretary in the DOE from 1975 to 1984, put it:

> the core of the impact which Informatics Technology can make on development is the *use of integrated computer-cum-telecommunication systems* to convert data into information and use/manipulate/control such information for planning and decision-making, efficiently managing large economic assets, enhancing productivity and safety, and improving the availability, quality and access to new services like telecommunication itself, railways, education and health. *It is in the capacity to design, develop, engineer, integrate, install, commission, support and maintain such IT-based systems, that India has built up an impressive capability.* [second emphasis mine].[22]

Some idea of the ambitious character of electronics R&D can be gained from looking at some of the programs undertaken. The TDC series of minicomputers pioneered by ECIL in the early 1970s was a national first but never became commercially successful, despite total protection and public sector purchase preference. The TTL 7400 series of logic chips developed by BEL in the 1970s and BEL's Liquid Crystal Display efforts suffered the same fate. On the other hand, computerized data-handling systems for the air defense network and the Army Radio Engineering Network, radar development efforts, solar photovoltaic cells by CEL, and, in the private sector, PSI Systems' R&D contract from a Japanese firm for the design of a 32-bit minicomputer (the Psirius), and CDOT's indigenously designed EPABX, RAX, and MAX have been successful on the whole.

Some very ambitious projects were Mettur Chemicals' indigenous development of polysilicon, the basic raw material for semiconductors, CDAC's (successful) efforts to develop by 1992 a parallel processing supercomputer capable of 1,000 megaflops speed, and Knowledge-Based Computer Systems. But the question that arises is: has the across-the-board and in-depth technological effort dictated by the objective of "total system capability" led to a frittering away of resources with a few successes here and there? What has been the cost of such a strategy? Why have Korea and Brazil achieved higher technological levels in the fields they have focused on in a shorter time-frame? Or has

> the country's R&D policy not been too successful because of lack of realism . . . the government has funded almost everything under the sun. We should have had more success if we had set our sights right, concentrated and conserved human resources. . . . In spite of our indifferent performance in innovating products that could be productionized and marketed, some R&D expertise and infrastructure has been created . . . the country is still a decade or two decades away in state-of-the-art R&D in electronic systems and components.[23]

The Spread of Applications of Electronics Downstream

The spread of applications of electronics downstream has accelerated from the 1980s on.[24] In broadcasting, the introduction of color transmission in 1982 and the special plan for covering 70% of the population by 1984, implemented since 1982 and still being extended, represented a quantum jump in applications. In telecommunications, a major changeover to electronic switching was initiated by the decision to produce digital switching equipment with CIT-Alcatel and CDOT technologies. Datacommunications are also spreading. The DOE's INDONET project for an intercity datacom network (mainly for corporate subscribers) became operational in 1986. NICNET, a nationwide datacom system for government, became operational in the late 1980s. Several other networks, including the DOE's Education and Research Network (ERNET) for academic datacom and various intraorganizational networks, have since become operational. In the defense sector, the most visible achievements have been the indigenously developed data-handling systems for the Air Defense Network and the Army Radio Engineering System. Indian industry is moving into fourth-generation distributed digital control systems especially in continuous flow process industries.

Computerization of banking has been plagued by labor-displacement disputes with the unions. In 1983, the Reserve Bank of India's Rangarajan Committee recommended the installation of ledger posting machines (basically PCs) in branch offices and larger systems in regional and head offices. Progress has been very slow in the decade that followed. The implementation of the Narasimham committee report on financial-sector reforms since 1992 should promote rapid computerization of banking, stock exchanges, and insurance in the post-1991 era of structural adjustment. On the whole, the spread of electronics applications remains tardy, constrained by finances as well as by the poor tele- and datacom infrastructure necessary for exploiting the full potential of information technology.

MAIN FEATURES OF INDIAN ELECTRONICS DEVELOPMENT, FROM A SURVEY OF THE EMPIRICAL RECORD

(1) In terms of output market size and exports, India has lost ground tremendously in electronics compared to Korea and even Brazil, despite starting roughly on a par with them and having a head start in having a dedicated policymaking agency.

(2) Indian electronics is overwhelmingly domestic-market-oriented and uncompetitive in price and quality. The emphasis has been on broad-based ISI, whatever the cost.

(3) The focus has been on the industrial electronic equipment segments right from the beginning rather than staged vertical integration backward into more complex products, beginning with consumer and component assembly. Even

within the industrial equipment segments, there has been little attempt to specialize in particular lines. Instead, there has been an attempt to produce a wide range of equipment indigenously, resulting in subcritical investments and inefficiency.

(4) As a matter of policy, the public sector has spearheaded Indian electronics development and until the 1980s dominated the scene. The largest firms are all in the public sector. Within the public sector, firms specialize in different products, there being no competition between them.

(5) The SSI sector, a uniquely Indian concept, has been promoted, especially in consumer electronics and components, and accounted, until recently, for the bulk of production in the former and a share in all but telecom and aerospace/defense. SSI operates at suboptimal scales and is generally inefficient. However, a range of products were reserved for SSI in the 1970s and many continue to be so.

(6) Foreign firms were severely restricted until 1991. They have faced entry and expansion barriers, or they have been allowed expansion only on condition of farming out production to SSI subcontractors even at the cost of efficiency—for example, Philips in radio receivers.[25] Foreign technology import was also restricted by royalty and duration restrictions, which limited the inflow of new technology and eroded the "quality" of deals struck by Indian firms.

(7) The domestic large-scale private sector has been limited, in effect, until the late 1980s, to computers, CIIE, and some components. It had been denied licenses to expand (or even enter consumer electronics) under the pro-SSI reservation policy and the MRTP Act of 1969.

(8) R&D, though not insignificant by NIC standards, has also been of an across-the-board, technologically intensive ISI type. Such a pattern has resulted in the fragmentation of resources and effort and in continuing technological lags and no commercial results, despite the acquisition of greater technological depth.

(9) The amount of investment in the industry is low in comparison to other countries; and even this has been fragmented by blanket ISI and SSI reservation, and to an extent by geographical dispersal. Even more important, complementary investments to create domestic demand for equipment and, derivatively, components, vital for an ISI strategy, such as in telecom and broadcasting expansion and computerization and in datacom infrastructure, have been inadequate and greatly delayed.

(10) Notwithstanding these features, considerable integrated system capability has been created for large-scale informatics projects in several key industrial sectors and in government operations—for example, the projects being implemented by NIC. Substantial technological depth and some competitive capability, as in software, have been acquired, laying the basis for future competitiveness in selected niches.

(11) Considerable liberalization has taken place in the 1980s, especially of freer input imports, but the basic problem of uncompetitiveness and technological lag remains, and the blanket ISI approach did not change significantly, de-

spite official recognition of the problems. The state's role has not shifted from direct producer and regulator to the "followership" role of facilitator in the manner of Korea, despite recognition of the need for such a shift for competitiveness, considerable deregulation, the growing share of the private sector, and the promotion of autonomous R&D centers such as CDOT and CDAC for technology development. However, the post-1991 period has seen fundamental policy changes.

THE POLITICAL–ECONOMIC CONTEXT OF INDIAN ELECTRONICS POLICY: THE OVERALL DEVELOPMENT STRATEGY AND INDUSTRIAL POLICY FRAMEWORK

Electronics policy functions within the overall framework of industrial policy constituted by trade policy, licensing (entry and exit) policies, foreign collaboration policy, industrial credit policy, taxation (especially indirect tax) policy, and public investment policies. The following points are noteworthy.

(1) Trade policy has been highly protectionist. Tariffs have been among the world's highest, effectively banning competing imports until the changes after mid-1991. There have been stiff duties on raw materials, components, and capital goods not locally made, raising the cost structure of domestic industry. A system of multiple lists—open general license (automatic import subject to duty), limited permissible list requiring import license, canalized list and a banned list—erected barriers. Foreign exchange clearances from the Reserve Bank of India (the central bank) and port clearance delays create further delay/barriers and raise costs.

(2) ISI was encouraged, but there were entry barriers for domestic firms due to the licensing policy constituted by the Industries (Development and Regulation) Act of 1951. All non-SSI manufacturing industry required a license from the Ministry of Industry that specified the capacity that could be set up. Expansion required a further license. Capacity ceilings were thus inherent in the licensing system and were at the discretion of politicians and bureaucrats.

(3) Two extensions to the licensing policy were critically important. One was the Monopolies and Restrictive Trade Practices (MRTP) Act of 1969. This defined MRTP houses as those with assets of over Rs 200 million (Rs 1,000 million since 1985) or more than one-quarter of the market in their line of business. Such "monopoly houses" required additional clearance over and above the usual industrial licensing procedure, under Sections 21 and 22 of the MRTP Act. Thus "monopolies" were defined as such before the fact, in contrast to, say, US antitrust legislation. The effect has been to hamper the growth of large industrial firms and conglomerates, and to promote overdiversification and suboptimal capacities. The aim has been to check the concentration of control over assets rather than checking monopolistic behavior, or promoting competition per se, or redistributing income. This is clear from the very definition of monopoly houses in terms of an arbitrary asset value bearing no

necessary relation to market shares or monopolistic behavior, or arbitrary licensed capacity percentages as an indicator of market share, again bearing no necessary relation to monopolistic behavior.[26]

(4) The Foreign Exchange Regulation Act (FERA) of 1973 defined foreign firms as those with over 40% foreign equity. Their entry or expansion into most areas was restricted, except for 100% export-oriented operations or when they brought in advanced technologies. Expansion, in effect, required dilution of foreign equity to under 40%, which could be achieved by expanding the domestic equity base, not necessarily by divestment. Consequently, FDI virtually stopped in India. In electronics, IBM, which dominated the installed base of computers, pulled out in 1978, after fruitless negotiations over the equity dilution issue. Philips, ICL, and Siemens remained, the first two agreeing to dilute foreign equity to under 40%. FERA was criticized as inane because an under 40% foreign holding can be perfectly compatible with foreign control if the rest of the stock is widely held. In India, due to de facto restrictions on takeovers in the stock market, it was and is common for controlling interests in firms to hold well below 40% of equity, because promoters' equity and government-owned financial institutions' stake typically made up a majority, the latter were pro-existing managements as a matter of policy, and only a part of the remaining widely held stock was actively traded.

(5) Technology import has been strictly regulated from 1970 to 1991. Royalty limits on foreign technology licensing agreements were limited to 5% of net value added—effectively 3% after 40% taxation of royalties. The duration of contracts was ordinarily limited to five years. Restrictive clauses such as bans on exports and tying of equipment supply were banned. The Indian Patent Act was amended in 1970 to give less protection to foreign patent holders. These restrictions made technology suppliers less ready to include anything beyond the minimum in the technology package transferred.

(6) The SSI sector was promoted from the late 1960s onwards with all manner of incentives, as it was considered labor-intensive and suitable for dispersal for regional development. This was not a venture-capital-type approach to promoting small but promising entrepreneurs, but a reservation of products for small firms defined by an arbitrary asset value. The reserved products were judged suitable for small-scale operations according to the technology of the day, reflecting a static view of technology and ignoring the possible importance of organizational economies of scale for competitiveness.[27]

(7) The Industrial Policy Resolution of 1956 reserved certain core sectors of the economy for 100% government ownership, including defense, aerospace, communications, and broadcasting. A vast public sector industrial base was created from 1956 on. The public sector was characterized by lack of exposure to foreign and domestic competition, many firms being monopolies and enjoying price preference in government purchases and characterized by inflexibilities including, in effect, permanent employment.

(8) After the nationalization of banks in 1969, the industrial credit policy reinforced the pro-SSI anti-large firm trend by reserving 40% of bank credit for a priority sector that included SSI in addition to agriculture and traditional artisanal industries. This was in sharp contrast to the Korean banking system's large, low-interest loans to the *chaebol*. Raising capital from the public via the stock market was also licensed in that all capital issues required authorization from the Finance Ministry, an inherently arbitrary bureaucratic decision.

(9) Indirect taxes have been the mainstay of government revenues. The main reason for this would appear to be that they are less visible and therefore politically easier to levy. Indirect taxes consist of customs duties, excise duties, state excise duties, and sales taxes levied variously by central and state governments. These taxes, being in percentages, have a cascading effect on costs and prices. A shift toward value-added taxes began only in 1986.

(10) Incentives were given for geographic dispersal and location in backward areas or compelled through the licensing system, especially for MRTP and FERA companies, which also had a cost-raising effect and hindered economies of agglomeration important in many industries including electronics.

Thus the overall industrial policy regime within which electronics policy has had to function has been one of blanket and intensive ISI, with an anti-big-firm, anti-foreign-capital bias, emphasizing deconcentration of economic power and geographic dispersal, ostensibly in the interest of equity, employment, and regional balance.

THE EVOLUTION OF INDIAN ELECTRONICS POLICY

The 1960s and 1970s: Blanket ISI for Self-Sufficiency

The government set up the Bhabha Committee on Electronics in August 1963, shortly after defeat in the India–China border war of October–November 1962. It submitted a comprehensive report in February 1966, a month after its chairman, nuclear physicist Homi J. Bhabha, the Chairman of the Atomic Energy Commission, died in a plane crash. It recommended that Indian electronics develop by going in for the latest technologies and become self-reliant as soon as possible, while recognizing the need for some time for foreign collaboration. Policy implementation was to be handled by the Electronics Committee of India, an interdepartmental committee dominated by the Defense Ministry's Department of Defense Supplies created in November 1965 to handle import-substitution and *inter alia* supply of electronics requirements following US aid (and electronics equipment) cutoffs (to both sides) during the India–Pakistan war of 1965. The Ministry of Defense also controlled BEL, then one of the two largest electronics enterprises. ECIL was created in 1967 by the DAE to specialize in nuclear instrumentation.

The Electronics Committee lacked the finances and staff to plan and implement an electronics policy. This was remedied by the setting up of the Department of

Electronics (DOE) in June 1970, reporting directly to Prime Minister Indira Gandhi. The Electronics Commission (EC) was created in February 1971. It was to be the policymaking body with the authority to establish public enterprises in electronics. The DOE was to be the EC's executive arm. The first Chairman of the EC and Secretary of the DOE was M. G. K. Menon, a solid-state physicist who had earlier been the director of the AEC's Tata Institute for Fundamental Research (TIFR) at Bombay. His principal aide was Ashok Parthasarathi. In October 1971, the EC formed a research group called the Information, Planning and Analysis Group (IPAG), which was to be the intelligence and planning unit. It was headed by N. Seshagiri, also formerly of the AEC's Tata Institute, who was to become the key figure in computer policy in the 1980s. The EC included the Cabinet Secretary and the Secretaries of the Finance Ministry and Planning Commission, as well as the Managing Director of ECIL, A. S. Rao. Initially the Defense Ministry and BEL did not have representation on the EC. The formation of the EC and DOE represented a victory for the electronics group in the AEC over the Defense Ministry.[28]

During the 1970s the EC/DOE attempted a public sector-led intensive and blanket ISI drive in manufacturing and technology development, with a focus on industrial electronic equipment and, to a lesser extent, on components.[29] The pro-public sector, pro-SSI, anti-foreign, and anti-large private sector, intensive and blanket ISI flavor of electronics policy under the new EC/DOE dispensation becomes evident in the following major policies and recommendations made during the 1970s.

The Semiconductor Committee Report of 1972 recommended mandatory installation of diffusion facilities for full vertical integration for semiconductor device production, despite uneconomically tiny domestic demand. The Minicomputer Panel Report of September 1973 and subsequent computer policy, including the promotion of ECIL's indigenously developed TDC series of minis, the freezing of IBM's computer installations after 1972, and the eventual exit of IBM in 1978 due to equity dilution disputes caused by FERA, reflected these policy orientations.[30] CMC was created in 1976 to maintain and support computer installations. There were inordinate delays in the late 1970s in granting licenses to the first emerging private computer manufacturers—DCM-Data Products, Hindustan Computers Limited, and ORG Systems—in an effort to protect ECIL. The Ten-Year Perspective Plan for Electronics and Communications of 1975 emphasized self-sufficiency and indigenously sustained growth within ten years, industrial dispersion, employment, and indigenous components and systems for defense.[31]

TV manufacturers in the SSI sector, numbering over 100 by the early 1980s, were licensed in very small capacities of 5,000–25,000 units. MNCs and large firms were denied expansion into TV manufacture, with Philips getting a TV license only in 1984. TV deflection components, tuners, and several consumer-grade components were reserved for SSI. Consumer electronics continued to be regarded as a luxury item, and efforts to reduce costs and expand usage were tardy. TV transmission coverage expanded very slowly to cover only 17% of the population by 1981.[32]

The SEEPZ was set up in Bombay in 1972 to attract MNC investment in labor-intensive assembly plants for exports of component subassemblies back to their home plants. This was an attempt to generate exports by the then increasingly common EPZ route. However, SEEPZ was none too successful. The report of the Menon Committee on Exports of 1978 went into the reasons for poor export performance, but its recommendations stayed within the parameters of the existing policy regime, including those features responsible for high costs, low quality, and uncompetitiveness, such as SSI reservation, licensing, and so forth.[33]

By 1978, it had become evident that the industry was not progressing as well as expected. The Janata government set up the Review Committee on Electronics (the Sondhi Committee) headed by the then Secretary, Ministry of Steel, Coal and Mines, Mantosh Sondhi. The Sondhi Committee's basic recommendations in September 1979 amounted to what was then a drastic liberalization package. It said that, "the essence of this package (of recommendations) is the dismantling of unnecessary controls, restrictions and regulations but with an overall coordinated and integrated approach on growth."[34] It recommended approvals by a single-point nodal agency and "substantial but selective investments in production-oriented research and development."[35] The package included a larger role for the organized private sector as against the public and SSI sectors. However, while criticizing the industry's achievements, the Committee kept within the existing policy framework, including the SSI policy. In 1980 the new Indira Gandhi government formally rejected most of the Committee's recommendations but, ironically, began a process of liberalization from 1981 on. This liberalization, uncoordinated and in fits and starts, gathered momentum and some sort of coherence after Rajiv Gandhi became prime minister in December 1984.

Liberalizing Policy Changes in the 1980s: Toward Efficient ISI?

The following major policies and developments took shape between 1981 and 1991:[36]

(1) The Components Policy of 1981 took note of the importance of scale economies and de-emphasized SSI.

(2) Color-TV transmission began in 1982, as did manufacturing of imported color-TV kits. The Industrial and Licensing Policy for Color-TV Receiver Sets of February 1983 removed capacity limits and led to entry by over 100 small assemblers. A crash program to expand TV broadcast coverage to 70% of the population by 1984 was launched, aided by the INSAT–1B satellite becoming operational in August 1983. Table 5.9 shows the expansion of TV broadcast coverage alongside the expansion of TV production.

(3) The private sector was allowed entry into manufacture of subscriber-end telecom equipment, such as EPABXs and telephones, in March 1984.

(4) The New Computer Policy of November 1984 ushered in the computer boom of 1985–90. A host of new entrants competed in a substantially deregulated market. Volumes soared and prices dropped sharply due to increased competition, lowered duties, and the fall in world peripherals prices. Manufacture remained dependent on imported components, and little R&D took place. However, the latest technology was introduced for the PC range based on the latest microprocessors.

(5) The Integrated Policy Measures on Electronics of March 1985 was a further consolidation of the liberalizations announced in an earlier package of August 1984. They included further duty rationalization, delicensing of components, office automation, and, conditionally, consumer electronics, exemption from MRTP, FERA, and backward area location policies, dereservation for SSI of many components, and freer import of technology. Also included were traditional indigenization and infrastructural programs such as the INDONET datacommunication network, LSI/VLSI chips in SCL, materials in the proposed National Silicon Facility, and so forth.

(6) The New Software Policy of December 1986 liberalized software import for stock and sale as well as for software export production. The latter was sought to be promoted with incentives, stiff export obligations, and penalties for nonperformance and incentives for foreign investors for 100% export-oriented projects. In 1990, the Software Technology Park scheme was initiated at several locations to promote software exports by foreign investors and domestic firms, inspired by the success of Texas Instruments operation since 1986. These were duty-free import areas subject to export obligations, which were provided with infrastructure, including core computer facilities and dedicated satellite links for software exports.

(7) In telecom, the Department of Posts and Telegraphs was split and a separate DOT created in 1986. A public sector company, the Mahanagar Telephone Nigam Limited (MTNL), was formed by a merger of the telephone departments of the cities of Bombay and Delhi. MTNL and the Videsh Sanchar Nigam Limited (VSNL), the international telecom company, were made public enterprises. No longer government agencies, they were legally enabled to raise capital from the market to finance expansion and modernization. In 1989, the Telecom Commission was formed, headed by the CDOT chief, Pitroda. The Telecom Commission centralized telecom policy in all aspects, preserving the government monolith, the objective being to give a big push for comprehensive modernization of the networks, the manufacturing base, and technology development, as well as for strategic negotiation with MNCs.[37]

However, this arrangement collapsed after the Congress party lost the November 1989 elections. The successor National Front government undercut the efforts of CDOT to develop digital main exchanges indigenously and resorted to imports of exchanges on the ground that CDOT was two years behind schedule and expansion of the network could not wait, given the waiting list for telephone connections. The real reasons involved departmental infighting and lobbying by MNCs to frustrate CDOT's efforts.[38] Since 1990, telecom policy has been in a state of transition, the drift being toward breaking up the DOT monopoly, separating service provision, policy, and regulation in line with international deregulatory trends in the industry and granting freer private and foreign entry into manufacturing, cellular telephony, radio-paging, other value-added services, and even basic telephone services.

(8) The role of the state began to shift, although to a very limited extent, from owner and regulator to infrastructure provider and technology supporter, the best examples being CDOT, CDAC, and the provision of datacom infrastructure for software exports.

(9) However, there was also a countertrend paralleling the liberalization of entry and imports, consisting of an intensification of ISI in advanced technology areas for strategic self-reliance. Thus several high-tech projects were launched with inadequate funding, including ECIL's efforts to indigenize an obsolescent Control Data mainframe computer. These trends repeated the mistakes of the 1970s efforts, such as fragmented and suboptimally scaled investments and a lack of complementary investments in computerization, broadcasting, and telecom expansion to create adequate demand for equipment and components. Indirect taxes on consumer electronics were not slashed; TV production declined absolutely after 1988, once the relatively high-income market had reached saturation—the only such decline in a developing country and one induced by faulty policy. And there was no harnessing of MNCs to promote exports.

The New Strategy since 1991 and the Impasse of the Mid-1990s

From July 1991 on, virtually all of electronics was opened up to MNCs, at least as joint ventures, in the new Congress government's IMF-supported stabilization and structural adjustment program. This led to the enunciation of a new strategic philosophy in electronics in 1992 that recognized the limits of the existing strategy but is also constrained by the uncompetitiveness of the industry so far created.[39] However, it was recognized that with an import intensity of production of 31% on average, the industry had to be able to export to be viable.

The new philosophy identifies the basic objectives to be: (1) the spread of electronic applications in all spheres to raise productivity and create domestic demand; (2) making industry globally competitive through strategic alliances for finance, technology, and markets; (3) self-sufficiency in selected key products and technologies for defense/aerospace applications, where a threat of technology export restrictions by the developed world exists. A distinction was made between those segments—such as process control—where the spread of applications was more important and those—like software and consumer electronics with export potential—where manufacturing was important.

For the first objective, a massive program of computerizing Indian industry and banking and expanding the telecom and broadcasting networks has been mooted. The latter will also create the infrastructure to support competitiveness and attract investment.

For the second objective, India has thrown open most of electronics to FDI on liberal terms and dereserved public sector preserves, and it proposes to use the traditional lever of large contracts for technology acquisition. The main strategy to access the world market and salvage the uncompetitive equipment manufacturing sector is seen to be contract manufacturing in alliance with MNCs. However, India's principal comparative advantage is seen to lie in software—where it is already the most successful developing-country exporter—and systems engineering. IBM's return to India in 1992 in a 50/50 joint software venture with Tata Information Systems Limited best symbolizes the new policy thrust.

The third objective of selective self-sufficiency is to be promoted by projects such as CDAC's PARAM parallel processing supercomputer, the reconstruction of SCL for key ASIC chips cross-subsidized eventually by contract design and manufacturing, and the like, to gain a technological base and possibly a market niche.

For achieving the second objective of competitiveness, two major schemes have been initiated. One of these is the Electronics Hardware Technology Park (EHTP) scheme along the lines of the Software Technology Parks, from September 1992 on. This is essentially a duty-free import area that can be entered by both foreign and domestic firms. Duty-free imports are not subject to any value-addition (VA) criterion, unlike in the existing—largely unsuccessful—EPZ and 100% Export-Oriented Unit (EOU) schemes. However, access to the domestic market at half the normal duty depends on at least 15% VA, defined as the percentage of net foreign exchange earned (difference between exports and foreign exchange outgo on imports) to exports. With 15–25% VA, access to the domestic market would be 25% of production for finished goods and 30% for components; with over 25% VA, access would be 30% for finished goods and 40% for components. EPZ/EOUs allowed only 25% access to the domestic market at the same rate of half the normal duty. Firms in the EHTPs had product flexibility in that they could alter their product and export mix. Domestic market access was not on a product-wise VA criterion but in aggregate, unlike in the EPZ/EOUs. It was hoped that the EHTP scheme would attract both foreign investors and domestic firms to ally for contract manufacturing, helping to make the Indian electronic hardware industry—both equipment and component—competitive.

The other major initiative was the promotion of software exports. The multi-location Software Technology Park scheme was consolidated into a Software Technology Park (India) autonomous society under the DOE, and its management was centralized. The basic idea was to increase the opportunities for Indian software exporters and foreign investors to develop offshore software—

that is, in India rather than on-site in the main markets (principally the United States) through the export of personnel. The latter was and is the prevailing pattern and has resulted not only in lower net foreign exchange earnings due to foreign expenses, but also a brain drain.

The bottleneck in doing offshore work was the inadequacy of international datacom facilities of adequate bandwidth and acceptable bit error rates. By providing dedicated satellite links, core computer and other infrastructure, as well as duty-free imports for software exports, it was hoped to increase offshore work and seize a bigger and more lucrative slice of the world software market. It was also hoped that the Indian software industry, concentrated as it is in coding and systems analysis and design (the fourth and second, respectively, in a five-category hierarchy of software activities in a declining order of complexity), could move into labor-intensive data entry and skill-intensive software maintenance, the fifth and third activities in the hierarchy. These require adequate datacom facilities for offshore work; if provided, software exports could potentially increase by orders of magnitude. The DOE took the initiative to provide as many as 90 out of 200 64-Kbps (kilobits/second) bandwidth datacom channels during 1992–94 to overcome the resistance and exorbitant rates charged by the DOT/VSNL.[40]

In telecom, as mentioned earlier, the drift since 1990 has been in the direction of freer private and foreign entry into manufacturing, value-added, and even basic telecom services, and of following the international trend toward breaking up the government monopoly and separating the functions of policymaking, regulation, and service provision. The National Telecom Policy of 1994 consolidated these trends but left intact the DOT's monopoly of policymaking, ownership of the two public sector service companies, MTNL and VSNL, and regulation. The proposed Telecom Regulatory Authority of India mooted in 1995 will be ultimately under the control of the Communications minister. Under such circumstances, both domestic and foreign private investment in the basic services opened up to the private sector will be deterred. An assuredly neutral regulatory agency is required. And if the targets set out in the Policy are to be met, massive private and foreign investment will be necessary. It is estimated that at Rs 32/$, over $17 billion will be required over 1994–97, of which DOT will have only $5 billion.[41]

Thus, despite the post-1991 policy changes, electronics policy and industry remain at an impasse. On the one hand, a new strategy has been crafted that marks a distinct shift in the role of the state toward privatization, globalization, and the "big followership" role of support and infrastructure provision, rather than regulation. In this connection, it must be mentioned that the state has tried to induce industry associations to come together on a consortium basis for policy formulation as well as exports, although without much success. However, for reasons of overall policy incoherence and lack of interagency coordination, and, in turn, inadequate datacom infrastructure, the EHTP scheme has not been notably successful in turning around the hardware industry, nor have software exports, though growing, really lived up to their potential. And the resources for meeting telecom targets are nowhere in sight. As for the third objective of self-reliance in critical components, SCL's wafer-fabrication facilities have not yet been rebuilt, having been put on hold since November 1994 due to lack of financial support from the key users, the Defense Ministry and the DAE.

ELECTRONICS POLICY OBJECTIVES: HAVE THEY BEEN MET?

The goals of India's electronics strategy can, broadly speaking, be summed up under the following six headings:

1. growth
2. self-reliance
   a. from imports
   b. from imports of technology
3. exports
4. employment generation
5. regional balance
6. promoting downstream applications and productivity

Have these goals been achieved? An unqualified yes is possible only in one sphere, that of regional balance. Plants have been set up in almost all states; most have their own public enterprises, and production has been dispersed.

As for employment, total direct employment in electronics has increased from 130,000 in 1981 to 300,000 in 1992.[42] This growth is marginal in the context of India's unemployment problem. Employment generation has been the rationale for licensing small capacities to large numbers of SSI units dispersed over the country in television, radio, other consumer items, and many components reserved for SSI.[43] The result has been the perpetuation of low-volume manual assembly operations for longer than would have been the case under unrestricted domestic competition. This has created high-cost, low-reliability and, hence, low-demand, low-volume production and low employment situations. It is unclear whether this has helped to create greater employment than would be the case if unrestricted domestic competition and/or export promotion led to a high-volume, low-cost, high-demand, high-volume virtuous cycle, in which employment levels were higher due to much higher output, despite higher capital–labor ratios, as in Korea. In any case, the technological trends toward capital-intensive automated production cannot be escaped. The really significant employment potential of the industry lies in the overall economic growth made possible by productivity increases downstream deriving from applications of electronics.

As for the export objective, it is clear from the figures that this has not been achieved; in fact, compared to other NICs—even import-substituting Brazil—India's performance has been very poor, with the singular exception of software. Growth, too, as an objective has been far below expectations. Looking at the record from the formation of the EC/DOE in 1971, India has lost ground relative to Korea, Brazil, and other developing countries.

As far as the spread of electronics applications is concerned, while there has been an acceleration of application spread since the early 1980s beginning with TV broadcasting expansion, digitalization of telecom switching, and general computerization, the pace has been very slow. Brazil and South Korea achieved computerization of the banking and financial services sectors much earlier and more comprehensively. Brazil has expanded its telecom network, digitalized switching, and introduced new services much more rapidly than has India. It has also expanded computer use in commerce and industry generally at a much faster pace, as has Korea. Korea has also automated more and more of its high-volume electronics assembly industries. One good overall indicator of the much more rapid spread of applications in Korea and Brazil is the faster growth of, and in absolute terms, much larger installed base in computers of every size class in value and in numbers as well as much higher telephone density.

Finally, what of self-reliance, the seemingly central and driving motivation behind the electronics policy regime? There are two aspects to self-reliance: (1) reducing dependence on imports, and (2) reducing dependence on imported technology and developing an indigenous technology-generating capability. As far as the first is concerned, self-reliance has increased, as measured by the declining ratio of imports to domestic production and consumption until 1990. Self-reliance has been achieved most in consumer electronics and least in components, especially ICs. As for technological self-reliance, a diversified R&D base and the integrated systems engineering capability for large industrial and infrastructural projects have been created. Some of these efforts, such as software, CDOT, and CDAC, are certainly successful, especially the RAX and the PARAM parallel computer, and are potentially competitive. But it is officially acknowledged in the post-1991 policies that the gap between Indian and world technology has been widening over the years and that there is an increased need for foreign technology. And the key to technological self-reliance—a design and manufacturing base in ICs, at least in ASICs—has not materialized.

Two points need to be noted about the self-reliance objective:

1. Heavy import-dependence in components and subsystems means that the high levels of import substitution achieved in equipment are basically import-dependent production, and in the final analysis also technologically dependent.
2. High import substitution and "self-reliance" can be achieved by producing obsolete equipment and obsolete technology (for example, in telecom switching, for many years based on obsolete strowger and crossbar technologies), in which case the "self-reliance" achieved is spurious, or at best dubious.

On the whole, one is forced to conclude that despite some undoubted successes, self-reliance is far from having been achieved, especially in the sense of earning a trade surplus or at least being able to export in selected specializations to pay for politically and strategically important if uneconomic projects in others. What has been happening is that the successive rounds of imported technology-dependent ISI have in each round failed to create competitive capability, eventually requiring a fresh round of imported technology and equipment.

NOTES

1. For the information in this paragraph, see Rajiv Rastogi et al., "Indian Electronics Industry: Geographical Spread," *Electronics: Information and Planning* (*EIP*), 21, no. 3 (December 1993). The brief overview of Indian electronics policy and development in this chapter is based on original sources, primarily the data of the Department of Electronics (DOE) and Department of Telecommunications (DOT), published in their Annual Reports and the DOE's journal, *Electronics: Information and Planning*, and on Annual Reports of the various electronics public and private sector enterprises; we also draw on the Indian electronics industry press, primarily *Dataquest*, and on material published by the various industry associations.

2. *EIP*, 15, no. 9 (June 1988): 594, Tables 1, 2; *EIP*, 15, no. 8 (May 1988): 526, Table 1. Conversions are at average exchange rates for the calendar year from IMF, *International Financial Statistics 1995*, pp. 436–437..

3. Computed from production figures in Tables 1 and 2.

4. For discussion of imports, see World Bank, "India-Development of the Electronics Industry-A Sector Report," Report Number 6781-IN, Vol. III, Annex 6, p. 9.

5. R. K. Verma, "Electronic Hardware Technology Park: A Step Towards Globalisation," *EIP*, 20, no. 6 (March 1993): 280, Table 4.

6. For an internal DOE assessment of India's potential market for various electronic products that based its analysis on per capita income and arrived at relatively pessimistic (or realistic?) conclusions especially as regards consumer electronics, see Ranajit Dhar et al. "Market Potential of Indian Electronics During the Seventh Plan 1985–86 to 1989–90," DOE (1985).

7. For the figures in this section, see *EIP*, 15, no. 9 (June 1988): 595; *EIP*, 21, no. 9 (June 1994).

8. For latest figure, see DOT Annual Reports, various years; DOT Annual Report 1994–95, p. 99.

9. Edquist and Jacobsson, "The Integrated Circuit Industries."

10. ITI, BEL, ECIL, BHEL Annual Reports.

11. "Report of the Working Group on Electronics Industry for the Eighth Five-Year Plan," *EIP*, 18, no. 2 (November 1990): 78; *EIP*, 18, no. 3 (December 1990): Table 15.17.

12. Organized = non-SSI private sector, domestic and foreign, and public sector.

13. *EIP*, 15, no. 9 (June 1988): 603, Table 6; "Report of the Working Group in Electronics Industry for the Eighth Five-Year Plan," *EIP*, 18, no. 2 (November 1990): 78.

14. See "Report of the Working Group on Electronics Industry for the Eighth Five-Year Plan," *EIP*, 18, no. 3 (December 1990), Table 15.17; *EIP*, 18, no. 2 (November 1990): 78.

15. "Report of the Working Group on Electronics Industry for the Eight Five-Year Plan," *EIP*, 18, no. 2 (November 1990): 75, Table 1.8.

16. DOE and DOT Annual Reports; *EIP*, various issues; World Bank, "India," Vol. II, p. 74, Table 3.2.

17. Press Note No. 10/25/86-LP, DOE (May 27, 1986).

18. DOE Annual Reports; World Bank, "India," Vol. III, Annex 15, p. 2, Table 1, and p. 25; Vol. II, p. 140.

19. World Bank, "India," Vol. II, p. 84; DOE Annual Reports, various.

20. DOE Annual Report, 1987–88, p. 8.

21. "Report of the Study Team on Science and Technology for the Eight Five-Year Plan for the Electronics Industry," *EIP*, 17, no. 10–11 (July–August 1990).

22. Ashok Parthasarathi (1987), "Informatics for Development: The Indian Experience," paper presented at the second session of the North–South Roundtable of the Society for International Development, p. 13.

23. Bureau of Industrial Costs and Prices (BICP), Ministry of Industry, "Report on Electronics" (1987), p. 84.

24. The following account of the spread of applications is based on various issues of *EIP*, DOE, DOT, and public sector firm Annual Reports, *Dataquest*, *Computers Today* and *Infotel* magazines, and press reports.

25. Interviews with Marketing Director, Philips, 22 January 1987, Bombay; and with Chairman and Managing Director, Asha Pavro (Pvt.) Ltd., 16 January 1987, Bombay.

26. For an elaboration from the Korean electronics case on the institutional (in this case, conglomerate form of organization with large constituent firms) basis of comparative advantage, see Mody, "Institutions and Dynamic Comparative Advantage."

27. For a stress on organizational economies of scale, see Mody, "Institutions and Dynamic Comparative Advantage." For some useful conceptual discussions of technology transfer, see World Bank, "India," Vol. II, p. 40; Chudnovsky, Nagao and Jacobsson, *Capital Goods Production*; Sanjaya Lall, "India," *World Development*, 12, no. 5/6 (May/June 1984),

28. The story of the politics behind these developments has been told in Joseph Grieco, *Between Dependency and Autonomy: India's Experience with the International Computer Industry* (Berkeley and Los Angeles, CA: University of California Press, 1984), ch. V, pp. 103–149.

29. The EC was wound up in May 1989. However, we use EC/DOE as a shorthand for electronics policymakers throughout the rest of this book.

30. For the text of the report, see *EIP*, 1, no. 5 (February 1974).

31. DOE, "Ten-Year Perspective Plan for Electronics and Communications" (1975).

32. *EIP*, 9, no. 11 (August 1982): 561.

33. For the text of the report, see *EIP*, 6, no. 2 (November 1978).

34. *EIP*, 7, no. 6 (March 1980): 388. For the text of the report, see *EIP*, 7, no. 6 (March 1980); *EIP*, 7, no. 7 (April 1980).

35. *EIP*, 7, no. 6 (March 1980): 388.

36. For the full text of the policy statements between 1981 and 1986 covering the first six points in this section, see *EIP*, 15, no. 1 (October 1987): 40–55. For the Software Technology Parks scheme see S. S. Oberoi, "Software Technology Park: Concept, Status and Procedure," *EIP*, 19, no. 3 (December 1991). For developments in telecom since 1986, see DOT Annual Reports since then.

37. Interview with Sam Pitroda, May 20, 1995.

38. See G. B. Meemamsi, *The CDOT Story* (New Delhi: Kedar Publications, 1994).

39. Interview, N. Vittal, Secretary, DOE, 18 December 1992.

40. Pronab Sen, "Software Exports from India: A Systemic Analysis," *EIP*, 22, no. 2 (November 1994); and interview, 25 May 1995.

41. Pronab Sen, "National Telecom Policy 1994: An Analysis," *EIP*, 22, no. 1 (October 1994).

42. DOE Annual Report 1987–88, p. 7; Rajiv Rastogi et al., "Indian Electronics Industry: Geographical Spread," *EIP*, 21, no. 3 (December 1993).

43. See all articles on the radio and TV industry in *EIP* in the 1970s and the Sondhi Report.

6

The Driving Forces of Indian Import-Substitution: The Political Economy of Indian Electronics Strategy

WHY DID INDIA'S STRATEGY FAIL EVEN IN TERMS OF ITS OWN OBJECTIVES?

If Indian electronics policy failed to achieve even its own objectives, let alone compare favorably with Korea, Brazil, and other NICs, three questions arise. (1) Why did such policies fail? (2) Why, then, was such an idiosyncratic ISI strategy chosen and carried out? (3) The Indian electronics policy experience in comparison with the Korean and Brazilian raises a number of vital questions about the strategic capacity of the Indian state to plan coherently and implement industrial policy and, following from this, about the concept of strategic capacity itself.

To answer the first question first, several interlocking factors were responsible for the failure of Indian policies. First of all, faster growth of the equipment segments even based only on the home market would have required radical reductions in cost and improvements in quality. Factors that raise costs, hamper quality, and impede competitiveness have been examined in two comprehensive reports in the late 1980s on competitiveness potential.[1]

The single most important cause of Indian electronics' lack of competitiveness has been the high cost of raw materials, components, and intermediate inputs caused by the indirect tax regime. For example, for a color television or a PC, the sales price, including indirect taxes, was 140% and 160% higher, respectively, than c.i.f. world prices.[2] For components, an electrolytic capacitor and double-sided PCB had sales prices 110% and 30%, respectively, above c.i.f. world prices.[3] Indirect taxes as a percentage of sales prices ranged from 26–33%, being

47% for color televisions. Customs duties and excise on imported components, materials, and capital goods accounted for 12–18% of sales prices; excise/sales taxes on final goods for 10–30%. Customs duties on imported materials and components alone accounted for 8–12% of sales prices. The duty on imported capital goods for new projects was 25%, and for expansion and modernization it was 55%.

The result of such an indirect tax regime was that prices of electronics goods were 50–150% above world prices.[4] Interest costs and profit levels were also relatively high compared to East Asia. Technology and infrastructure were additional factors in raising costs. Underlying these cost differences were policy-induced firm-level characteristics such as lack of economies of scale at plant and firm levels and obsolescent technology, reinforced by lack of domestic competitive pressure. A pro-SSI licensing and credit policy, MRTP, and FERA, had resulted in a proliferation of suboptimally sized firms. Trade protection and entry barriers due to licensing and public sector reservation had reduced domestic competitive pressures. Combined with technology import restrictions, these had resulted in obsolete technology.

Among policy-induced firm-level characteristics, a lack of economies of scale was the most important in reducing competitiveness, especially in the capital-intensive components segment. Here fixed costs of equipment, such as wave-soldering and flowsoldering machines and automatic test equipment (needed also in computers and consumer electronics assembly), are very high and become economic only at high volumes.[5] A pro-SSI policy in consumer electronics and many reserved components inhibited automated soldering, perpetuating manual assembly. The latter results in low quality and reliability, hence inhibiting demand. Automated assembly becomes technologically necessary for color television and more complex PCBs with more chips.[6]

Organizational economies of scale are also critically important in assembly industries where there are scale economies in marketing, after-sales service, and R&D. Furthermore, firm size enhances negotiating power in acquisition of technology.[7] These firm-level economies of scale could not be reaped because of the policy bias against size. The best example that one can think of is the damage done to possible competitiveness in radio receivers since the late 1970s, despite a large home market because of the MRTP/FERA bias against Philips. Philips was forced to subcontract its production to 11 small-scale manual assemblers, when it could have moved to large-scale automated assembly, reaping significant technological and organizational scale economies as well as cost reductions and quality improvements. This would have both expanded domestic use and possibly made it profitable for Philips to export from India.[8]

The policy regime, therefore, perpetuated small manual assembly operations in consumer electronics and kept costs high and quality low, thereby constraining demand growth. This, in turn, affected the demand for components, especially discrete semiconductors and linear ICs, which needed higher volumes to be economic. There was a contradiction in the policy regime between consumer and component segments, and between both of these and the lack of complementary investments to boost equipment demand such as expansion of broadcast coverage and computerization.

Questions and Issues That Require Explanation

Why did Indian policy choose a broadfront ISI strategy beginning with industrial electronics rather than a staged vertical integration backward starting from simpler component and consumer electronics assembly? Given such a choice and capital and home-market constraints, why was there not more selective entry and more gradual vertical integration in tune with output growth for scale economies? Why was there not a more pragmatic attitude to foreign firms to access technology and world markets, given the small home market, scale economies, and rapid technological change? Why was there a protected public sector monopoly or predominance in many segments of industrial electronics, neglecting the need for competition, well into the 1980s? Why were consumer electronics and components reserved largely for SSI well into the 1980s, excluding rather than harnessing large private firms and MNCs for exports on OEM or joint venture bases as possible entry strategies to the world market? This could have earned foreign exchange to finance firm-level R&D as well as cross-subsidize uneconomic but politically prioritized developments?

Given an ISI strategy, why were the large-scale complementary investments in telecom and broadcasting expansion and computerization not made? And why was domestic demand expansion discouraged by high customs duties, even on imports not made locally, and by high excise duties on final products, particularly consumer electronics—especially since a components base was considered basic for self-reliance, and component production depends on equipment segment growth and is scale-sensitive. Lastly, but crucially, the policy flexibility dimension—why was Indian policy, with its dedicated policy institutions, unable to make appropriate course corrections in the 1980s, even after the shortcomings of the existing policy framework and the need for competitiveness were recognized?

WHY DID INDIAN ELECTRONICS POLICY EVOLVE THE WAY IT DID? AN ANALYTICAL FRAMEWORK

If Indian electronics, despite a specially created EC and DOE, did not achieve most of its objectives and if the reasons for not doing so were the policies made and the conditioning framework, why, then, was such an idiosyncratic strategy chosen? The question must be dealt with at two levels of policy in order to provide an adequate answer: (1) at the level of electronics policy, both for each segment and for coherence across segments; (2) at the level of the overall industrial and economic policy regime. In a number of cases, electronics policy may be directly determined/constrained (determined by default) by overall policy; in others, electronics policy may have had considerable freedom of choice.

Following Haggard and Cheng, Adler, Evans, Johnson, Deyo, and Nayar, we can attempt to explain strategy by specifying three categories of variables:[9]

(1) *The ideologies/thinking of policymakers in electronics and industrial strategy generally.* This category has been given explanatory primacy by Adler in his study of Brazilian computer strategy and Argentinian nuclear strategy. It assumes a very high level of autonomy of the policymakers from social forces, leaving their mindset as the key factor.

(2) *The interests of the constituents of the dominant social coalition, including both state elites and dominant social groups.* Haggard and Cheng, Johnson, and others emphasize the more-than-just-relative autonomy of the state elite, clearly distinguishing between its own interests and private capitalist interests. These may be political, not just economic. For India this has been stressed by Bardhan in explaining the tendency toward nondevelopment expenditures (subsidies, etc.) in crowding out productive investment. There is also a large neoclassical literature on rent-seeking activity hampering economic growth.[10] And following Evans in his explanation of Brazilian computer strategy—stressing interests over Adler's ideology—the interests of different sections of capital are taken to be key influences on state policy. This type of explanation has to do with the social—coalitional basis of development strategy and the position of the state elite in the coalition.

(3) *The influence of the "high politics" of national security and alliances (in India's case nonalignment) in emphasizing self-sufficiency regardless of cost.* This is very often a neglected dimension as underlined by Evans and by Haggard and Cheng.[11] We shall argue that this is an indispensable element in explaining Indian electronics strategy.

We shall argue here that these three sets of explanatory variables operate at each of the two levels of analysis mentioned—that is, directly at the electronics policy level and indirectly through their impact on the overall industrial policy framework.

A crucial characteristic in interpreting the evolution of electronics policy at both levels under the influence of combinations of the three variables is *policy flexibility*. While Nayar attributes the path-dependency of policy flexibility to the interests created within and outside the state by past policy, we would go a step further to include *ideology too as a path-dependent constraint on policy flexibility*. The path-dependency of policy flexibility is also, we would argue, due to the ideological legacy inherited by the state leadership. Past political and ideological positions taken by the leadership can be significant constraints to policy flexibility especially if such positions are perceived to underpin the legitimacy and popularity of the leadership and if actions and statements contradicting such positions may undermine the leadership.

We first try to explain electronics policy at its own level of analysis with the help of the three categories of variables outlined and the policy flexibility dimension. Failing a satisfactory or complete explanation, we move back into the level of overall industrial policy and try to explain its impact, direct or indirect, on electronics policy, again employing all three causal variables and the policy flexibility dimension.

ENTRY, SELECTIVITY, AND PHASING: WHY BLANKET IMPORT-SUBSTITUTION?

The question of why India chose to enter and vigorously promote industrial electronic equipment even before a simpler consumer electronics and component assembly industry was well-established—the diametrical opposite of what was done by Korea and other East Asian NICs—needs explanation. Given that the Indian market was very limited and India far behind world technologies, it would appear impossible for Indian firms to develop competitive capabilities in these more complex products.

The lack of Korean-style phasing of entry into complex and scale-dependent products requires explanation, as does the lack of selectivity. Indian policies on entry, phasing, and lack of specialization have been severely critiqued, among others, by Mody, the World Bank, and the BICP, all from an economic rationality standpoint.[12] Sanjaya Lall's critique of India's broader technology policies also falls into this category.[13] His point when extended to electronics implies that while premature vertical integration backward into advanced products and technologies can promote precocious scientific and technological development, such advances cannot be commercialized on the world market because the domestic market does not permit the scale economies and revenues—and, hence, recurring R&D expenditures—needed to compete in industrial equipment and advanced components. ISI carried too far into advanced equipment and R&D activity creates a foothold but does not allow a steady climb and saddles the industry with inappropriate plant and technology for its stage of development. Thus, the social costs incurred by protection of technological learning processes do not ultimately pay off in competitiveness. Overprotection of technological learning misdirects efforts other than in remunerative directions such as mastery of production knowhow. Yet we find that from their inception the EC/DOE promoted indigenous manufacturing of industrial equipment.

The Causes of Such Policy Choices

The "High Politics" of National Security

The causes of such policy choices lie at both the electronics and overall policy levels. At the electronics policy level, they encompass ideology, interest, and national security considerations. For the ideology and national security variables, explanations at the electronics policy level and the overall industrial policy level dovetail. Industrial policy influenced electronics policy both directly and indirectly.

At the level of "high politics" of national security, foreign policy, and overall political–economic strategy, a number of key developments formed the backdrop to the formation of the EC/DOE. The October–November 1962 border war with China led to an awareness of the need for military modernization. The 1965 war with Pakistan reinforced this and heightened the sense of threat. There was now a distinct possibility of a two-front war in the future, given the unresolved disputes with both Pakistan and China and growing closeness between the two. In addition, China went nuclear, testing its first bomb successfully in October 1964. This put intense pressure on India to follow suit or at least not to forswear the nuclear option.

When the Nonproliferation Treaty was signed in July 1968 by the United States, the Soviet Union, and the United Kingdom, followed by a host of other states, India refused to join in, keeping open its options about a future nuclear-device-testing program. It was also making the complementary industrial self-sufficiency moves, most notably the setting up of ECIL in 1967 under the DAE for nuclear electronic instruments and the work of the Bhabha Atomic Research Centre (BARC) Electronics Group, which later resulted in ECIL's indigenously designed TDC minicomputer series. These moves signified a greater drive toward autarky in the nuclear—including electronics—sphere, as a hedge against the possible sanctions that might be triggered by a nuclear test.[14] The policy direction set by such autarkic pressures in BARC, DAE, and ECIL carried over into electronics policy in the EC/DOE because the atomic energy policy group (Rao, Menon, Parthasarathi, Seshagiri) came to rule the roost in the new policy institution. It would most certainly have been accentuated after the Indo–Soviet Treaty and the Indo–Pakistan war of 1971 had strained Indo–US relations. The latter development, when seen in context of the NPT having just come into force in 1970, increased the possibility of technological sanctions.

Another important background event was the beginning of a Sino–US rapprochement following US Secretary of State Henry Kissinger's secret visit to China via Pakistan in July 1971. The impact of this on India's strategic situation was felt immediately in the later part of 1971. The newness of the situation was highlighted by the US tilt toward Pakistan despite Chinese diplomatic support of Pakistan and the possibility of Chinese intervention on Pakistan's side in the event of an Indo–Pakistan war. This situation was quite unlike the US support for India in 1962 and neutrality in 1965, and despite Washington's official view of China as a hostile power.

What was new about the situation was that whereas during the 1960s China had been on bad terms with both superpowers, and hence India could depend upon support from either or both without being under pressure to go nuclear herself, this was no longer the case. The implicit guarantee provided by the Indo–Soviet Treaty could only grow less credible over the years against a China with a growing nuclear arsenal and US ties. Therefore by 1971–72 pressures to go nuclear were undoubtedly stepped up from within the national security establishment. But this would inevitably provoke technological—including electronic—sanctions. Hence the pressure to design and manufacture a broad range of industrial electronic equipment including nuclear instrumentation, computers, telecom equipment, as well as critical components, was bound to have mounted. The people in command in the new EC/DOE would have to have been keenly aware of these requirements.[15]

The "High Politics" of Macroeconomic Strategy in Response to the Crisis of ISI

The mid- to late 1960s were also a period of profound domestic political and economic crisis.[16] The famines and food imports of 1965–67 precipitated a foreign exchange crisis. The devaluation of the rupee in 1966 as a result of World Bank pressure made imports more expensive without boosting exports, as is frequently the case in developing countries where imports are inelastic while local import-dependent industry cannot respond quickly to devaluation.

One possible response to the crisis would have been a broad reorientation away from ISI toward export orientation, as many East Asian countries and other now-NICs did at the time, taking advantage of the boom in world trade and wage–cost advantages. This did not occur, despite pressure from the World Bank. Fundamentally, the political base for such a shift did not exist. Quite to the contrary—the major interests in the dominant coalition had a stake in the state-controlled ISI model. By the late 1960s, ISI had created a substantial large as well as small-scale protection-dependent private industrial sector. The state elite, both politicians and bureaucrats, had strong vested interests in maintaining their discretionary powers in industrial and import licensing and foreign exchange allocation. The licensing system operated in a way that shielded established or early entrants in an industry—usually the big firms—from competition from new entrants. Therefore, at least two key constituents of the dominant coalition, industrial capital and the political–bureaucratic state elite, had vested interests in the state-regulated ISI model.

The organized labor movement of all party affiliations and the political and intellectual left—the Left proper as well as the left wing of the Congress party—were also strongly opposed to any such shift, which would have been seen as a sellout to foreign capital under World Bank and US pressure. Whatever tendency there may have been toward an export-orientation solution to the foreign exchange crisis ended with the defeat of its potential political base in the Congress party split of 1969. The potentially proliberalization organization (right) wing of the Congress party was defeated in the split. The Indira Gandhi-led Congress (Indira) that emerged victorious, supported by the Left from outside, included the pro-ISI and pro-public sector left wing of the party, many of whom gained important positions in the cabinet and inner circle of the Prime Minister. Thus the

Congress party and the government leadership that emerged shared a populist pro-ISI ideology that was hostile to the private sector and foreign capital, and pro-public- and small-scale-sector.

The nationalization of 14 major banks, most of which were closely tied to big industrial conglomerates, and the passage of the MRTP Act in 1969 constituted a defeat for the large-scale private corporate sector and for the possibility of a reorientation toward export orientation, since small firms would require even greater protection and state support. The nationalization of foreign oil companies and the passage of FERA in 1973 further consolidated the public sector-led ISI model.

By the end of 1969, it was clear that the solution to the foreign exchange crisis (implicitly the crisis of the first round of ISI) was to be a more intensive second round of ISI. This was, in fact, already being implemented since the 1966 devaluation, when the payments imbalance was controlled by reducing imports. To this was added FERA and restrictions on technology imports and royalty payments to save on outgo of profits, dividends, royalties, and technical fees. The 1973 oil crisis and the resulting pressure on foreign exchange gave further impetus to ISI.

Such was the "high political" and national security backdrop to electronics strategy from the early 1970s on. The national security variable influenced electronics policy in an ISI direction on these issues, directly as well as indirectly through the effect of the overall policy regime created from 1969 on. Both these effects, direct and indirect, only reinforced the direct effects of the ideology and interests variables at the level of electronics policy itself. Therefore, an interlocking and mutually reinforcing set of causal factors operating at policy levels set in motion an intensive and blanket ISI orientation in electronics policy, emphasizing autonomous capability in industrial electronic equipment. There were no countervailing political or ideological forces at either policy level.

WHY THE NEGLECT OF INTERNATIONAL COMPETITIVENESS?

If national security and economic autonomy were the prime objectives of the new strategy, then why did it neglect international competitiveness? For if these were the objectives, one would expect electronics policymakers to recognize that in a fast-changing technology the achievement and maintenance of autonomy would unavoidably require finances for the recurring import of technology, adaptation, and R&D. This, in turn, would require the industry to be able to finance itself at least partially by earning foreign exchange. Even if a self-sufficiency-oriented ISI strategy was politically necessary, at least some parts of the industry would have to be internationally competitive in order to sustain politically motivated but uneconomic development programs in other product categories. *For if nothing was going to be competitive, then technological self-reliance requiring recurring innovation in a fast-changing world could never be economically sustained, eventually aborting the very goal of self-reliance for national security and economic and political autonomy. Without economic self-reliance there can be no technological self-reliance in the long run. Hence, international competitiveness should have defined itself as a goal at least for some segments of the industry in the eyes of the policymakers.*

It is necessary to emphasize that the crisis of the first round of ISI did not pose a stark choice between a second round of intensified, blanket ISI and all-out economic liberalization. Nor should the previous paragraphs be construed as advocacy of the latter option for the time. It was perfectly possible to conceive of an intermediate "efficient ISI" strategy that selectively pursued ISI in sectors vital to national security and economic autonomy with export orientation in selected potentially competitive industries, with an appropriate mix of public and private enterprise including selective foreign collaboration, combining the use of market forces for generating competitive pressures for efficiency within a framework of overall state guidance.

Why Indian strategy never defined international competitiveness as a goal can best be understood by looking into what this would have involved. For the Indian state to be able to implement an international competitiveness-oriented "efficient" ISI strategy, it would not only need to be insulated and autonomous enough and possess the machinery to be *able* to do so; it would have to have a *political interest in defining its objective to be such.* This brings us back to the political situation at the time of the setting up of electronics policy institutions.

The overall industrial policy regime constraining electronics policy took shape for purely political reasons as part of Mrs. Gandhi's political strategy during 1969–71, and continuing into the 1970s. It was not as if there had been no alternative that the Indian state could have followed. As mentioned earlier, an "efficient ISI" strategy, including a degree of export orientation, was possible. The Fourth Five-Year Plan draft embodied such a vision. The clash between this document and the Ten-Point Plan of the Congress party embodying a socialist approach, including nationalization and a greater role for the public sector, was precisely what provoked a split over policy within the party in April 1969.[17]

By this time it had become clear that the party-organizational leadership, for reasons beyond the scope of this study, was maneuvering to replace Mrs. Gandhi as prime minister. Mrs. Gandhi's strategy from then on was to ally herself with the party's left wing on policy issues in order to weaken, isolate, and defeat her enemies in the party-organizational leadership, projecting them as reactionaries and herself as a socialist. Essentially, a struggle for power was turned into an ideological struggle over policy direction in order to defeat a threat to her position as prime minister. In the process, a series of populist, specifically anticapitalist, pro-public sector, pro-SSI, and more regulatory anticapitalist policy measures were promulgated. The new round of ISI was to be characterized by several uniquely Indian features, with profound implications for electronics policy.

Mrs. Gandhi continued with such policies, including several nationalizations and FERA, even after winning a two-thirds parliamentary majority in the 1971 elections. The elections had been won on a populist slogan of removing poverty.

Even after this it remained very important to maintain her ideological credibility and public image as a socialist, not just within the party, but towards the nation in general. For this had become the main political plank and means of rallying popular support for Mrs. Gandhi and her faction of the Congress party—the Congress (Indira), as it came to be called.

What is crucial here is that the economic–strategic choices made at this time were motivated primarily by political–strategic imperatives—a subordination of the economic to the political in policy matters. This was not a loss of insulation due to penetration of the state by private interests, as in Brazil. It was a subordination deliberately imposed by the prime minister in a battle for political survival. It does not therefore represent a lack of strategic capacity due to a lack of political closure in Deyo's sense.

This subordination of the economic to the political was present in the choice of intensified ISI rather than export-oriented liberalization as the response to the foreign exchange crisis. The same logic applied to the choice of other policies whose unintended effect was to impede international competitiveness. The battle fought by Mrs. Gandhi against the party-organizational leadership in the split of 1969 and deliberately turned by her into a seemingly left-versus-right ideological struggle for the party's general policy direction constrained her government's freedom of choice in economic policy. For, having chosen to depend on the left within and outside the party and fight elections on a populist platform, she was constrained to follow policies that would not contradict the socialist image created.

The result was not simply populist sloganeering about removing poverty, but hard-to-reverse measures such as bank nationalization and the MRTP Act. Later in the early 1970s, the populist line continued with the passage of FERA and the reservation of 40% of bank credit for priority sectors, which included small-scale industry. Import-substitution in technology was also part of the general policy thrust. Restrictions on foreign technical collaboration were instituted and patent protection for foreign patent holders reduced in the Indian Patent Act of 1970. These policies were more political than economic. As blows dealt to her opponents, they were extremely effective. The big business houses and foreign firms whom she suspected of supporting her opponents in the party and in the right-wing Swatantra party were put on the defensive and made dependent on the state for credit, and they had to clear additional licensing hurdles under MRTP and FERA. Mrs. Gandhi emerged as a nationalist and pro-people leader opposing the reactionary forces of big business and foreign capital.

These policy measures had extremely important implications for the possibility of selective international competitiveness-oriented policies even within ISI. They formed the framework of the industrial policy regime within which electronics policy had to function. To the then existing industrial policy regime of import and industrial licensing, reservation of industries for the public sector, and so forth they added new restrictions against large private and foreign firms.

Within such a framework it was impossible to promote international competitiveness with its implications of liberalization, competitive pressures, and picking winners for state support.

Such a path was chosen not because of the lack of political closure and, hence, strategic capacity of the Indian state. Rather, it was part of a deliberate political strategy. The question became one of strategic priorities at the leadership level rather than lack of strategic capacity at the overall industrial or electronics policy level. The state was unwilling (rather than unable) to impose long-term national priorities over the immediate political priorities. The prime consideration at that time was the fact that such a strategy would threaten to undermine the government's socialist and nationalist credentials. It would make it vulnerable to attack by key support groups within and outside the party for being pro-capitalist, anti-public sector, anti-self-reliance, selling out to MNCs, and so forth.

Within such an overall regime constraint, even "efficient ISI" could not be defined as an objective at the electronics policy level. It was only when the cumulative effects of this neglect of efficiency via the balance of payments and growing need to borrow began to be felt in the aggregate at the national level in the 1980s that talk of "efficient ISI," competitiveness, and exports began.

However, a great deal still remains to be explained. For even within a broad ISI orientation, a great many policies could have been moderated and more could have been achieved. Indian ISI was characterized by several rigidities. They included an extremely inflexible attitude to FDI, a bias in favor of SSI (a uniquely Indian concept) and public enterprise and against large private firms, high indirect taxes, and (given that an ISI strategy was adopted) a lack of adequate complementary investments to boost domestic demand. In the following sections we shall examine the causes of each of these policies.

THE ATTITUDE TO FDI

Indian electronics policy toward foreign capital and technology could have been more flexible without necessarily allowing foreign domination. Korean electronics has had far larger FDI, not only in 100% foreign-owned semiconductor assembly operations, but also in import-substituting and export-promoting joint ventures with the Korean *chaebol* (and involving transfer of technology on terms superior to India's).[18] Yet this has not stifled Korean electronics technology, initially behind India's but now ahead in semiconductors, computers, and other areas. Korean electronics exports have been overwhelmingly from Korean firms. Brazilian telecom is another case in point. Despite domination of equipment production by MNCs, Brazil's CPqD had designed a series of digital exchanges long before India's CDOT started operations in 1984 and has been installing them since the early 1980s.

These experiences pose the question of whether within the same overall ISI strategy there could have been more accommodation to foreign capital to achieve greater inflow of new technology and access to world markets on an OEM basis. However, FERA applied across the board. While it was theoretically possible to make discretionary exceptions within the FERA guidelines for high-technology and export-oriented industries, such exceptions were *not* made by the DOE in electronics. These microlevel decisions need closer examination.

The three most important cases of the application of FERA in electronics are those of IBM and ICL in computers and of Philips in consumer electronics. The IBM case has been examined in detail by Grieco.[19] We go over it in brief here to compare it with Brazil and with Indian treatment of the other two MNCs. The government restricted all new computer installations by IBM in 1972. After FERA, IBM, and ICL were asked to dilute all foreign equity to below 40%. IBM refused to comply, at least in its main line of business—mainframe computers, installation and maintenance. At the time IBM was only refurbishing and selling used computers; the size of the market did not justify manufacture. However, in 1972 it offered to manufacture locally the obsolescing IBM–370. The government, however, kept up the pressure to dilute. In April 1976, IBM made what was up to that point its best offer. It offered to export line printers and software from India, share equity in its sideline operations (card manufacture and data services), and set up a scientific center built around an IBM–370/145 and an electronics testing facility. It also offered not to compete with ECIL minis then on the market, the TDC–312 and 316, which were superior to the IBM–360/30's and 360/40's that the company offered to refurbish and install in 1972. In return, IBM wanted exemptions from FERA for its hardware manufacturing and maintenance operations. IBM's offer was finally rejected by the Janata government in July 1977. Subsequently, in November 1977, IBM announced its decision to leave India, and it pulled out by June 1978.

ICL and Burroughs were willing to meet the government's FERA conditions. ICL agreed to merge its two (manufacturing and sales) units and dilute foreign equity to below 40%. Burroughs agreed to set up a 100% export-oriented joint venture with Tatas on a 50/50 basis in SEEPZ. These decisions were made before IBM's final offer, its rejection, and the consequent decision to quit. However, ICL continued to retain control of the newly formed International Computers Indian Manufacture (ICIM), even with under 40% foreign equity, as was all too common in India. ICL continued therefore to make key decisions in its Indian operation, which remained a domestic-market-oriented one. However, it was forced to bring in an advanced mini, the ICL 2904, in 1978 so as not to threaten ECIL's TDC–316.

Grieco is of the opinion that on balance India lost out by being too tough and forcing out IBM. He claims that its last offer would have been advantageous for India and, realistically speaking, the best that India could have bargained for at the time. The principal loss, according to his assessment, was the noninitiation of line printer and software exports, which would have generated the foreign exchange to pay for imports of computer systems. Such imports increased from the late 1970s on, as ECIL could not meet domestic requirements.

In any case, IBM's promise not to compete with ECIL meant that the market could have been segmented; IBM could have been kept to its large systems and national firms to minis and micros. As the market grew, IBM could possibly have been induced to manufacture large systems locally. It could also have been induced to export software and peripherals from India on a large scale, as it did in Brazil and elsewhere and as Burroughs did in India. Brazil launched its market reserve policy for national firms in minis and micros in 1977, keeping IBM confined to large systems and calculator exports, rejecting its offer of locally manufacturing the System–32 mini, just at the time when IBM was quitting India. IBM's presence in no way hampered the growth of local computer firms. And, perhaps, had such a strategy been followed, up-to-date IBM mainframes could have been made in India, rather than the essentially obsolete Cyber 180/810 and 830 contracted for by ECIL in 1986.

One can add that dilution to under 40% did not lead to any lessening of ICL's control over ICIM. In the hypothetical case of dilution by high-technology IBM, the arbitrary 40% mark would have been as meaningless as it was in ICL's case. Control could have been exercised even with minority equity by control of technological upgrading.

Furthermore, de facto ICL control of ICIM did not hamper the rapid growth of local firms after 1984. By 1987, ICIM had lost its leading position in sales ranking to Hindustan Computers Limited and was unable to prevent the emergence of competition, including ECIL, in the supermini range. The presence of foreign capital need not necessarily stifle indigenous firms if the latter are allowed to collaborate freely with other MNCs as has been the case since 1985. Despite an ICL-controlled ICIM, the latest machines from then on were made under technical tie-ups with MNCs superior to ICL in the world market.

On the other hand, former and incumbent senior policymakers in the DOE continued in 1987 to maintain that the IBM rejection was the right course. IBM was blamed for being shortsighted; they should have stayed on even on FERA terms for future market expansion. One senior official attributed IBM's obstinacy to the personality of the then IBM chief in India, A. L. Taylor, who clashed with DOE Secretary M. G. K. Menon repeatedly over the IBM–360. He claimed that IBM tacitly recognized its mistake by later demoting and then firing him. However, this does not answer the questions posed by the Brazilian experience. The key to IBM's quite different behavior there would appear to be the size of the Brazilian installed base and market growth and the fact that it did not have to dilute equity.

The best explanation for the push for indigenization in computers is the need for self-reliance for national security.[20] The computer initiative at ECIL originated with the BARC Computer Group, the precursor to ECIL's Computer Division.[21] It was headed by A. S. Rao, later managing director of ECIL and EC member. S. Sreekantan started the TDC–12 program at Trombay. In 1970, while on a visit to Japan with Vikram Sarabhai, then AEC chairman, Rao argued that not enough and only uncoordinated work was being done in computers. Rao, along with Sreekantan and R. Narasimhan of BARC, submitted a report advocating a computer development effort. They estimated a total requirement of Rs 21.30 million for the TDC–12, 16, and 32, including software development. The first meeting of the EC gave the entire money to Rao at ECIL, half as grant, half as loan. The TDC–12 and 16 were on the market by 1972–73.

The ECIL effort eventually ran out of steam only because of lack of DOE support. The DOE was blamed for allowing liberal imports of Digital Equipment Corporation minicomputers—35 over 1973–75, 200 by the end of the decade. However, former ECIL chief executives had to admit, as industry watchers have commented, that ECIL failed to develop a marketing strategy and therefore failed to compete with the new private entrants of the late 1970s. It therefore remained more or less confined to the preferential-purchase public sector market. And, as Grieco has pointed out, the TDC series was too far behind imported minis on price–performance criteria to be attractive to users, even given all the delays attendant on imports. A former ECIL chief executive blamed the high cost of peripherals and the import duty structure.

However, to return to the question of the attitude to foreign capital, it is not at all clear that IBM's presence, and even ICL's 2904 after 1978, were in any way the cause of ECIL's failure. Their machines were not competing in the same range as ECIL's. The partitioning of the market as in Brazil would still have left a protected minis and micros market for domestic firms. Therefore, it would appear that the reasons for the exclusion of IBM were essentially ideological, at the overall industrial policy level as argued by Grieco. At the electronics policy level, they would seem to have been motivated both by ideology and ECIL's interests. A former ECIL chief executive admitted and took credit for actively fighting IBM. ECIL probably wanted to eliminate IBM despite its not competing with ECIL machines, out of fear of a future threat materializing should an IBM offer to make minis in India become attractive to the government at some point.

This explanation of rigidity on FERA need not have applied outside the computer field if national security reasons were the main reasons behind the strong stand on FERA. What of major cases where national security considerations or an EC/DOE infant like ECIL did not exist? The case of the only other major MNC operating in India—Philips in consumer electronics and components—is instructive. Philips had been present in India for over four decades. It manufactured, among other things, radios (since 1948) and electronic components (since 1959) and exported radios from India. Philips was compelled to dilute to 39.7% under FERA; it complied by 1980.[22] The question that arises is whether it would have been possible to use Philips for exporting electronic components and radios from India on a large scale in the 1970s when assembly activity was labor-intensive in exchange for domestic expansion licenses and exemption from FERA? (The World Bank in 1987 suggested such a strategy for the future in selected products.)

Philips' exports from India grew from a mere $1 million (1971) to $5 million (1985). As a percentage of sales, its exports remained very low: 1% (1971), 6%

(1976), and 2% (1985). Thus its growth from $63 million (1971) to $210 million (1985) was almost wholly domestic-market-based. Philips is a global producer of consumer electronic ICs, but it chose not to develop India as an export base. In 1985 it exported only 200,000 radios from India out of sales of 1.787 million and production (at its own plants excluding SSI subcontractors) of 994,000.[23]

It appears that Philips did make commitments to export components from India in exchange for licenses to expand there and promised to consider exporting radios on a large scale in discussions during 1976–78. According to EC/DOE sources, Philips reneged on five commitments to export components and never came up with radio proposals. According to Philips, they were exporting some radios from India to pay for their imports under the foreign exchange entitlement scheme, but exports were not really profitable as compared to operations in Taiwan, where by 1985 they were exporting 120 million ICs.[24] This was so even at SEEPZ because of poor infrastructure and Indian labor not being competitive with East Asian labor. Therefore it made no economic sense for Philips to export from India.

In addition, BEL did not want Philips to make ICs in India, as it feared leakage from export-oriented production to the domestic market at a time when it was hoping to establish itself as a chipmaker through its TTL 7400 bipolar series.[25] As for exports from the DTA, high material costs and the Indian policy of forcing subcontracting to SSI did not allow automated assembly or economies of scale. Such automation and scale economies were becoming essential for efficient radio manufacture after the introduction of transistors and then ICs.

Philips was not granted exemption from FERA and had to dilute foreign equity. However, while Philips retained control with minority equity, it was treated as a separate entity by Philips of the Netherlands, thus hindering its (and India's) access to Philips technology and exposing the inanity of FERA. It would appear that infrastructural and other overall policy-induced deficiencies made it impossible (even if they wanted to) for electronics policymakers to exert pressure through licensing and FERA on the company to induce major exports by it of radios and components from India. In fact, from May 1985 on, restrictions against foreign-majority companies in components, materials, and closely held high-technology products were lifted even for the domestic market, but they showed no interest, preferring to license out technology, because such manufacture on a tiny scale was not found worthwhile in India.

The IBM, ICL, Tata Burroughs, and Philips cases lead to two conclusions: (1) A more liberal attitude to foreign equity would probably have led to a greater transfer of technology to Indian subsidiaries and joint ventures than was the case. This view is also supported by studies on transfer of technology to India from Europe.[26] But even more critical than the extent of equity participation allowed would be the competitive pressure felt by subsidiaries, joint ventures, and collaborators in India, and Indian policy toward technology imports.[27] This is illustrated by the inflow of technology to Indian partners since the New Computer Policy of 1984. (2) As far as using FERA exemptions and licensing of domestic expansion as instruments to induce MNCs to export from India is concerned, the basic problem was that the cost-raising overall policy regime and poor infrastructure made India uncompetitive as a location. Consequently, though MNCs may have done (Philips), or offered to do (IBM) some exporting, this would have been under pressure and even then would not have matched their Far Eastern operations. Furthermore, the view was strongly expressed by two senior former policymakers in the EC/DOE and the public sector that, for essentially political and strategic reasons, even with infrastructural and incentive improvements MNCs would show no interest in India as compared to "client states" in the US orbit like Korea and Taiwan. They would not have been willing to locate or subcontract assembly production in a nonaligned country friendly with the Soviet Union when they had politically less risky options.[28]

In retrospect, it appears that at the electronics policy level policymakers did not seriously consider the option of selective world-market orientation in certain labor-intensive products even within an overall ISI strategy, for essentially political and ideological reasons. And they considered impossible the necessary changes that such a policy would require for domestic and geopolitical reasons. We have detailed the political reasons at the overall policy regime level. The priorities and vested interests of the overall regime also reinforced electronics policy and ruled out making the policy changes needed for a partially export-oriented strategy to work. These are clear cases of not only ideological determinants operating at both electronics and overall policy levels, but also of path-determined ideological constraints on policy flexibility, whereby certain possibilities were not seriously considered because of their political–ideological unviability.

WHY SMALL-SCALE INDUSTRY AND THE PUBLIC SECTOR?

Another important—indeed, unique—way in which Indian ISI differed from Brazil's or Korea's was the way in which the private sector was divided into the SSI sector and the "organized" (non-SSI) private sector, as large- and medium-scale industry is called. Within the organized sector, there are further restrictions on the so-called "monopoly" houses—large industrial firms falling under the purview of the MRTP Act. Electronic consumer products and many components were essentially reserved for SSI until 1981 (for most of the reserved components) and 1985 (for most consumer products). Why were such policies carried out, given their negative effects on efficiency and competitiveness?

The Bias in Favor of Small-Scale Industry

The pro-SSI bias of overall industrial policy had an impact on electronics policy from the beginning. The EC/DOE started functioning in 1971 just after the bank nationalization and the reorientation of industrial credit toward priority sectors, including agriculture, cottage, and small-scale industry. At the same time, the MRTP Act curbed the expansion of "monopoly" houses, and licensing policy began to discriminate against them. An ever-growing list of products in each industry was reserved for SSI.

In electronics policy, the first pro-SSI action was the licensing of small firms for TV receiver production in extremely small capacities of 2,500–25,000 units. The decision to promote TV production by SSI was based on DOE's Information, Planning and Analysis Group (IPAG) Analysis Report 2 authored by N. Seshagiri, which noted at the outset that "it is of utmost importance for a country like India to plan for the proper level of automation, taking into consideration the abundant supply of skilled manpower and critical foreign exchange position."[29]

The report concluded that no

> economic advantage results from increasing the scale of production beyond n = 80 . . . no single unit or explicit consortia (should) be given a license exceeding 20,000 per year . . . safe range for licensing may be placed between 5000 and 20,000 per year. . . . The higher limit or thereabouts can be allowed for medium scale private industries and public sector electronics units including various units of the state governments . . . the lower limits or thereabouts . . . for small-scale private sector industries.[30]

The report totally ignored organizational economies of scale derived from discounts on bulk purchasing of components and parts. It also totally ignored the trend toward the use of transistors (later ICs) as against valves—that is, toward fully solid-state sets. The implication of such a trend was that manual assembly operations became inefficient because of their effect on quality and reliability, and hence on yield, and on competitiveness in quality. These problems would be solved only by a move toward flowsoldering (partial automation of assembly), requiring expensive equipment whose installation would become economical only at vastly higher volumes. Capacities of the order of 2,500–25,000 in the 1970s condemned the industry to a low-volume/high-cost/low-technology/low-quality uncompetitiveness trap. This situation persisted until the post-1982 TV boom. Meanwhile Japan had begun to automate assembly in 1970, and Taiwan, Hong Kong, and Korea by 1974–75. In India, the move toward flowsoldering began only in 1982–84, even on the part of the then largest manufacturers, such as Weston and Dyanora. Automatic testing equipment was not installed by these two firms until 1984 and automatic insertion equipment only in 1988 and 1987, respectively.[31] Even these developments were due to technological compulsions. The higher volumes necessitated by the switchover to fully solid-state, even for b&w televisions, by 1980 and the use of ICs rather than transistors in color televisions necessitated, in turn, flowsoldering, automatic testing, and, increasingly, automatic insertion.[32]

The story of radio receivers and TV components (such as deflection components and tuners) licensed preferentially to, or reserved for, the SSI sector is similar. Even as late as 1986, there were as many as 100 TV assemblers, the largest having production volumes only on the order of 200,000–300,000, even allowing for the ownership of several SSI firms by one person/family (Table 6.1).[33] Korean volumes at the firm level were of the order of millions.

The important point to be noted here is that this pro-SSI policy in consumer electronics and components originated in the EC/DOE itself. The report quoted above was authored by an official who was the chief of the DOE's thinktank, IPAG, and who became the principal promoter of liberalization in computer policy. DOE officials basically limited their thinking to the parameters of the overall regime rather than fighting it, even though that framework ruled out competitiveness in consumer electronics—something that should have been clear to industry-watchers in the 1970s.

A DOE report elaborating on the Bhatt Committee Report on Small and Medium Entrepreneurs of 1973 states that "labor-intensive products include consumer goods, test and measuring instruments and other equipment which essentially require assembly-like operations for production."[34] It adds that "Electronics is best suited for enterprise development in backward areas."[35] The EC's recommendation for reservation of eight products was endorsed by the report.[36] IPAG reports throughout the 1970s reiterated that electronics, especially consumer electronics assembly, was a labor-intensive industry suitable for employment generation and geographic dispersal. Even the Sondhi Committee Report paid lip service to the role of SSI.[37]

Thus the damage—recognized as such in the 1980s—was done from within the EC/DOE independently of the overall regime's bias despite an expert staff. The evidence is that the overall regime's bias was tacitly absorbed by electronics policymakers and not questioned or opposed on techno–economic grounds. As one former senior policymaker put it, they operated within the broad economic and social philosophy of the government.[38] The pro-SSI policy bias can therefore

Table 6.1: Distribution of TV Companies in Terms of Production Volume

	Black & White TV sets			*Color TV sets*		
Annual Production	*1982*	*1985*	*1989*	*1982*	*1985*	*1989*
Up to 500	49	39	40	17	31	46
501–2500	23	42	34	9	32	24
2,501–5,000	15	15	20	5	16	12
5,001–10,000	7	18	23	3	10	8
10,001–50,000	15	28	31	—	20	27
50,001–100,000	1	9	9	—	9	2
100,001–200,000	—	—	4	—	—	1

Source: Electronics Information and Planning, Vol. 18, No. 11 (August 1991), Tables 6 and 8, pp. 609–610.

be laid down to serious analytical errors, combined with an intellectual inertia that passively absorbed the ideological biases of the overall policy regime without attempting to combat them—another clear case of path-determined ideological constraints on policy flexibility.

The pro-SSI bias was also reinforced by the SSI interests created by the path historically followed by electronics policy. In fact, SSI found it profitable to remain "small" to avail of various concessions. This led to industrialists fragmenting operations into several "small" companies to bypass the regulations, which only further militated against economies of scale at firm and production-run levels and against the upgrading of technology and quality.

In addition, the SSI electronics sector organized itself into a formidable lobby, especially in consumer electronics, but also in components. This lobby made it extremely difficult for the EC/DOE to reverse policy and liberalize entry for large firms, even after it was recognized that SSI reservation was a techno–economic error. Two SSI-dominated industry associations emerged in the 1970s and became effective lobbies in the 1980s. The Indian Television Manufacturers' Association (ITMA) was founded in 1974 (renamed Consumer Electronics and Television Manufacturers Association—CETMA—in 1993) and the Electronic Component Industries Association (ELCINA) in 1967. ELCINA's membership includes not only SSI firms but also BEL, ECIL, ITI, and other public sector giants.[39] However, its president and 14-member executive committee have always been dominated by the private sector, which in the early 1980s was mainly SSI. Although from as early as 1976 the DOE took note of the fact that components were by and large not suitable for SSI and sought to promote large-scale units and discourage SSI, the New Components Policy materialized only in 1981.[40] It took the EC/DOE five years of lobbying and internal struggles to defeat the SSI lobby in the industry and the political support it had received from the Janata/Lok Dal government of 1977–79.[41]

Some indications of the entrenched mindset in favor of SSI at both the electronics and general levels were the Menon Committee on Exports of 1978 and the Sondhi Committee Report of 1979. Despite the importance of the organizational—and for components, technological—economies of scale, the Menon Committee continued to pay lip service to SSI and avoid a direct attack.[42] The Sondhi Committee, although calling for the liberalization of licensing for the organized sector and taking note of scale economies, continued to repeat the myth of the suitability of SSI for electronics. In fact, the Committee chairman Mantosh Sondhi admitted that the report's recommendations criticizing SSI and favoring liberalized entry for the organized sector were deliberately toned down so that they would stand a chance of getting a serious hearing at the political level in the Janata/Lok Dal government—another clear case of the prevailing ideology as a constraint on policy flexibility.[43]

By 1981, however, when the new components policy was announced, ELCINA, dominated by the larger private firms, did not protest officially, because existing firms would not be hurt. It went along with the policy and called for more liberalization, persuading its members that there was room for all to grow.[44] But it was only in 1985 that consumer-grade components such as TV deflection components and tuners were dereserved and the components industry was delicensed. Many SSI members of ELCINA had private reservations about this but did not protest.[45] According to one source in ITMA, ELCINA could not protest because they had nothing to show in terms of price (net of taxes) or quality in SSI firms.[46] The important point to note here is that the ideological mindset, inertia, and vested interests at both the electronics and overall policy levels combined to delay reform of components policy until 1981 and in important products affecting consumer electronics till 1985, during a period of extremely rapid expansion worldwide.

In consumer electronics, the Indian Television Manufacturers Association (ITMA), which by 1986 had over 115 members, was an overwhelmingly SSI-dominated organization, militantly against large and foreign firms.[47] ITMA became active in 1982. The new policy on color-TV receivers in 1982 led to the entry of a large number of small assemblers. ITMA at that time fought very hard to exclude MRTP and FERA companies and foreign brand names, singling out Philips, which it saw as the greatest competitive threat.[48] Philips had been denied a TV license since the 1970s and was granted one for b&w sets only in December 1984.[49] ITMA also lobbied successfully against the extension of delicensing of consumer electronics in 1985 to MRTP companies. In fact, it managed to limit delicensing to only those companies not borrowing from public financial institutions, thus seemingly trying to pre-empt big firms from setting up large automated plants. Thus the SSI lobby successfully delayed policy changes that would have enabled the industry to operate at viable scales, introduce new technology, reduce costs, and raise quality. ITMA lobbied at both the EC/DOE level and the overall level—that is, the Ministry of Industry and the Development Commissioner for Small-Scale Industries. Over the years, SSI in electronics and in Indian industry in general had become a powerful lobby that the ruling party could not ignore, accounting for a significant share of output and employment.

The Bias in Favor of the Public Sector and Against "Monopoly" Houses

The bias against the "organized" private sector, particularly MRTP and FERA companies, originated as a result of the influence of the overall policy regime after 1969. At the electronics policy level the organized private sector—especially MRTP firms—were discriminated against because of a strong pro-public sector bias rather than a pro-SSI bias.[50] This bias held that the development of the favored industrial equipment segments be best left to the public sector.

The rationale for the public sector spearheading the thrust derived from the general pro-public sector ideology and from the usual economic rationale for public intervention. Public investment in manufacturing is usually rationalized as necessary in the cases of market failure due to externalities, lumpiness of investments, and uncertainty (due to long gestation, among other things). Technological externalities, lumpiness, and long gestation would certainly apply in the case of semiconductors and some advanced products. But it is doubtful that they would apply across the board to the extent that the public sector bias seemed to imply.

Indeed, it can be questioned whether such market failures occurred at all, since the private sector was not given a chance. Toward the end of the 1970s, once the new private computer manufacturers had been licensed (with some restrictions), they rapidly overtook ECIL in minicomputers, as pointed out by Grieco and admitted by a former ECIL chief executive.[51] There was a feeling that the private sector would not invest in R&D and would not be able to absorb technology. Indeed, it was felt that they would not even be interested in doing so, finding more profitable the soft option of successive rounds of technology import, which would defeat the national objective of technological self-reliance.[52]

This argument at the electronics policy level mirrored that at the overall industrial policy level. It was also reinforced by entrenched interests in the public sector, particularly ECIL and BEL, both of which feared private sector competition. Thus ECIL took credit for lobbying the exit of IBM and tried hard to restrict new entrants in computers, with considerable success, in the late 1970s.[53] BEL, led by C. R. Subramaniam, who also authored the Semiconductor Committee Report of 1972, wanted to impose the diffusion stage on all semiconductor manufacturers rather than permit assembly of diffused wafers. This was ostensibly to promote self-reliance, but also to protect BEL, which was doing the capital-intensive diffusion step, from competition from low-cost private assemblers. This resulted in several private assemblers, such as Greaves Semiconductors, Semiconductors Limited, Ramtronics, and others, being saddled with heavy fixed costs.[54]

It never seems to have been asked whether such private sector behavior was due to the trade- and licensing-policy-induced lack of competitive pressures, and whether more innovative behavior could have been induced by allowing greater domestic competition and selective foreign competition in electronics. Rather, it seems to have taken the overall policy regime and ideology as given. It also appears to have indirectly been a product of the macropolitical strategy of self-sufficiency based on short-term import-avoidance at all costs to conserve foreign exchange and for national security.[55] Thus path-determined ideological constraints were a major cause of the policy bias.

WHY THE LACK OF COMPLEMENTARY DEMAND-SIDE POLICIES?

Given an ISI strategy and a small domestic market in a populous but poor country, one would have expected to see complementary policies that boosted domestic demand for electronics in the aid of economies of scale and learning.

Such complementary policies would have been critical for the efficiency of the ISI model. Large complementary investments in broadcasting expansion, telecom modernization and expansion, and general computerization to create high demand and economies of scale for the equipment segment and, indirectly, for components were a logical requirement of an ISI strategy. So was the rationalization of the duty structure to lower the costs and prices of electronic goods and expand domestic demand. Why did Indian policy, unlike the massive expansion of telecom, computerization, and banking automation in Korea and Brazil, not effect complementary demand-side policies?

As far as complementary investments are concerned, the rapid expansion of the telecom network did not take place until the early 1990s for the following reasons. Apart from the lack of resources, there was the problem of the bottleneck in switching equipment. Rapid expansion could only proceed with digitalization of switching and with modification of the imported crossbar technology to Indian conditions of extremely high usage per line. However, if one compares India to Brazil, the digitalization decision and its implementation were painfully slow. Lack of adequate expansion of broadcast coverage was, firstly, due to the fact that it was not seen as a priority area, given the tremendous competing demands on resources. And, secondly, it would have been far less cost-effective to go in for a terrestrial network in the 1970s, before the planned communication satellite. Even after INSAT–1B was planned and expansion became easier, the decision to expand TV coverage greatly was essentially a political decision taken in 1982 by Information and Broadcasting Minister Vasant Sathe and Mrs. Gandhi, with the next elections in mind.[56]

As for computerization, the problem was one of general bureaucratic inertia, combined with lack of competitive pressure on managements in industry and banking. There was also the problem of very powerful resistance from the bank unions, which managed to delay the implementation of the Rangarajan Committee report (1982). This was quite unlike the case in Brazilian banking, which was substantially private and part of industrial conglomerates and which computerized quickly, starting in the late 1970s. As far as installation of process-control equipment in industry was concerned, it was partly due to "efforts to develop indigenous technology, which until recently tended to block local producers from access to developments in other countries."[57] And partly "it also reflects the limited awareness and interest of industrial users," undoubtedly because of the lack of competitive pressures, especially in public sector heavy industries.[58] The lack of complementary investments was due to factors largely beyond the EC/DOE's control—namely, due to the lack of resources, the perception of priorities at the level of overall planning and the lack of competitive pressure in the domestic market, labor union resistance in the key banking sector, governmental inertia, and three years of infighting and instability during the Janata period.

The question also arises as to why the customs and excise duty structure remained so cost-raising throughout the 1970s. Significant rationalization of the duty structure for electronics inputs did not begin until 1983. Until then, there were very high duties on even capital goods, raw materials, and components not locally made. Beginning in August 1983, import duties on materials and components were significantly lowered. But it was not until 1986, with the introduction of MODVAT (Modified Value-Added Tax), that indirect taxes became deductible through successive manufacturing stages. As a World Bank report put it,

> the effects of indirect taxes have no doubt encouraged vertical integration by electronics firms and contributed to insufficient horizontal specialization. Finally, the resulting higher prices paid by final consumers of electronic goods depressed demand for goods with relatively high price elasticity, for example, consumer goods, which, in turn, inhibited the production of components.[59]

How did a dedicated policymaking agency allow such a state of affairs to persist? The answer is quite simple as far as customs duties are concerned. The EC/DOE has no power to set indirect taxes. The Ministry of Finance was always concerned with revenue losses in the short term rather than with the cost structure of industry.[60] The tendency toward increased reliance on indirect taxes persisted because they were easier to collect, and politically easier to levy, because they were less visible to the public. In the case of high duties on materials and components, this was partly the EC/DOE's fault, arising from its penchant for unphased, premature, and cost-ineffective ISI in components. The manifest failure of the overall strategy, combined with the activism of some officials in the DOE (for example, Seshagiri in computers), led to duty rationalization from 1983 on, subsequently getting top-level political backing in 1985.

THE POLITICS OF LIBERALIZATION OF ELECTRONICS POLICY IN THE 1980s

Delayed Liberalization of Computer and Telecommunications Policies

Other important events (or nonevents) that require explanation are the long delay, from 1978 to 1984, in articulating a computer policy, and the delay in the implementation of the digitalization of telephone exchanges. In 1978, after IBM's exit and ICL's equity dilution, the government licensed four new entrants to the computer industry after intense lobbying, despite ECIL's obstructionist efforts. They were HCL, ORG Systems, DCM Data Products, and IDM/Nelco. It imposed limits on capacity, total turnover, size of system (restricting them to micros to protect ECIL's TDC minis), and import of peripherals and components. Yet these companies grew faster than ECIL and introduced new micros competing with ECIL. It was only in November 1984 that the EC/DOE finally liberalized computer policy on entry and input imports.

A computer boom followed. But the five or six years from 1978–79 to 1984 were lost years. These were the years when the micro emerged on the world scene, making entry by newcomers possible in a big way. It was during these years that Brazil's market reserve policy became operational and its national firms took off. It was also during these years that Korean firms went from nothing to exports of the order of millions in terminals, monitors, and microcomputers, even their domestic microcomputer market overtaking India's. Neither national firm dominance nor growth and technology development took place until the early 1980s, the price performance gap being rapidly closed only after 1982 and especially after 1985.[61] Why were these years lost?

Grieco asserts that in 1978–80 the government stalled on liberalizing entry for newcomer firms so as to protect ECIL, the EC/DOE's favorite. Liberalization followed only after the exit in December 1978 of M.G.K. Menon as Secretary, DOE, and Chairman, EC, and his replacement by B. Nag, a liberalizer, and even then only after intense lobbying by the new entrants.[62] The politics of liberalization of computer policy consists of several strands, as far as we have been able to uncover.[63] First, the Sondhi Committee Report (of which Nag was one of the coauthors) recommended a liberalized policy package, including greater private sector entry in computers. The Report, although largely rejected by the incoming Indira Gandhi government in 1980, had an impact within the EC/DOE in its emphasis on scale, costs, and prices, despite strong opposition from public sector advocates within. The new Secretary, DOE, from 1981 until 1984, P.P. Gupta, also tended then to resist liberalization, though he later became a dynamic chief executive of software exporter CMC.

The key proponent of liberalization in the DOE at this point was N. Seshagiri, earlier an advocate of SSI in consumer electronics. Now, apparently as a result of watching world trends and the industry's poor performance, he had become a convert to liberalization. For three years he carried on an internal struggle in the DOE, promoting measures eventually contained in the new policy. He took over as Additional Secretary, DOE, in charge of computers, control, and instrumentation in November 1982. It took him one year to clean up some of the irrationalities in the duty structure and some of the associated controls in the existing policy. He pushed for further decontrol but was opposed from above and by others against liberalization. While waiting for the transition from P.P. Gupta to S. R. Vijayakar as Secretary, DOE, he started moving on his own, bypassing his bosses and sending the first draft of the New Computer Policy direct to the Prime Minister and the Deputy Minister for Electronics, M. S. Sanjeevi Rao, himself an electronics engineer. The latter was supportive.

The existing private sector computer manufacturers, who had organized themselves into the Manufacturers Association for Information Technology (MAIT) by December 1983, were not behind the liberalization policy. They feared competition, and Seshagiri did not bring them in to help him lobby, afraid that they would scuttle his plans. Certain pro-liberalization individuals were taken into his confidence. In addition, Seshagiri, in his capacity as Director General of the National Informatics Center (NIC), the organization in the DOE responsible for promoting computerization in government, drew up application plans for all departments. In this he was opposed by the Directorate General of Technical Development and by general bureaucratic inertia. However, the "lucky coincidence" of Rajiv Gandhi becoming Prime Minister helped clear the roadblocks. It is very likely that Seshagiri, who was in charge of coordinating electronic scoreboards and networking for the Asian Games in Delhi in 1982, came into close contact with Rajiv Gandhi, then All India Congress Committee General Secretary and also coordinator for the Games, who reportedly took great interest in the electronic systems being installed. It was also through close contact with Rajiv Gandhi over the Asian Games that Philips, a major contractor, influenced him in order to enter TV production.

Seshagiri's lone battle for liberalization received support when S. R. Vijayakar, managing director of ECIL, took over as Secretary of the DOE in May 1984.[64] Vijayakar had seen the ECIL TDC series, including its latest, the TDC 332, fail to hold its own. The ECIL computer division had never shown a profit. He was in favor of selective ISI and a pragmatic attitude to foreign technology and capital. Eventually he came to question the viability of ECIL's mainframe computer indigenization efforts and India's semiconductor thrust (the TDC 332 had been forced to use BEL's chips). He had also seen ECIL's efforts at Data Acquisition Systems fail, despite two years of strong support and total protection. Vijayakar eventually came to support the World Bank report's criticisms of Indian electronics. He thought "deepening" had gone too far too early. He had canceled, in 1982, ECIL's indigenous color-TV development efforts, concluding that it would have cost at least 20% more than the kit-assembled sets. He came to the DOE convinced that building volume production based on cheap imported inputs was a necessary step before "deepening" into components, peripherals, and advanced equipment. He blamed the five-year delay in computer policy on the ISI ideology, general inertia, the Janata regime's pro-SSI stance, and the fact that Menon and Nag, being from a purely R&D background, were not business-oriented. They did not appreciate the importance of volume on costs, prices, expansion, and eventually on innovation itself.

Thus the critical five-year delay in formulating a new computer policy was caused by many factors, mainly at the electronics policy level, but also reinforced at the overall industrial policy level by ideological biases in favor of ISI. The liberalization of computer policy, too, originated within the DOE itself and was the result of a complex internal struggle over policy. The advocates of change emerged victorious due to support at the topmost level—that is, after Vijayakar, the Secretary, DOE, and the prime minister were convinced, probably helped by Rajiv Gandhi's contacts with Seshagiri—not due to private sector pressures or part of a general reorientation of industrial policy.

In telecom, too, there appears to be a pattern of policy incoherence and missed opportunities. India had a head start in the existence of ITI, with its production of electromechanical crossbar exchanges, and the Telecommunication Research Centre's (TRC) indigenous design of a small (1,000-line) electronic exchange, the SPC–1, which was fabricated by ITI and BEL and had completed field trials by 1979. The TRC had actually begun an ESS development project in 1965 and completed a 100-line lab-scale model by 1974. Yet India made the "go digital" decision only in the DOE-Posts and Telegraph (later DOT) Department Interdepartmental Committee's report in 1979 (which recommended two ESS plants of 500,000 lines each with foreign collaboration to be phased in comparatively slowly over four years), and the DOT's Sarin Committee on Telecommunications' 1981 report.[65] In the meantime a lot of time and money was wasted in trying to modify the inherently unsuitable Pentaconta crossbar exchange to suit Indian conditions.

The main factor behind the opening of peripheral equipment manufacture to the private sector in March 1984 was the growing waiting list for telephone connections. The public sector ITI and the DOT were unable to produce and install enough lines to meet the growing demand. In 1975, there were 637,000 on the waiting list for new connections, compared to 1.3 million lines installed. This situation was expected to become worse through the 1980s. The average waiting period for a telephone connection had grown to several years![66] The main bottleneck was the outdated central public switching technology used in main exchanges. This was either the obsolete strowger technology or the Indian modification of the ITT Pentaconta crossbar technology originally licensed in 1964.[67] There was a pressing need to switch over to digital exchanges.

It was for this that ITI signed a contract with CIT-Alcatel of France for its E–10-B digital exchanges in 1982 after a couple of years of negotiations. This was two years after Brazil's contract with Ericsson for transfer of similar technology to Ericsson do Brasil, by then Brazilian-controlled, despite an indigenization policy dating only to 1974. The result was that the DOT feared it would not have the resources to make complementary expansions in the peripheral equipment end and would have to resort to imports. The induction of the private sector into the manufacture of peripheral equipment was partly to raise the required investment funds and avoid import dependence and partly as a result of the influence of the US-returned expert Satyen G. "Sam" Pitroda.[68] The formation of the two public sector corporations, MTNL and VSNL, in 1986 was to enable them to float bonds to raise money from the public legally. The setting up of CDOT in 1984 was also in keeping with the indigenization drive. It was entirely due to Pitroda's tireless lobbying since 1981 and his promise of time-bound development of indigenously designed modular and upgradable family of digital exchanges.[69] These developments were not part of any coordinated plan.

In contrast, Brazil, which gained national ownership over telecom services only in 1968, set up TELEBRAS only in 1972 and CPqD only in 1976, and moved to gain national control over telecom equipment manufacturing only after 1975, made its digitalization strategy decision as early as 1976. The Ericsson deal for the AXE–10 ESS was negotiated by 1980 and implemented by 1982. Other MNCs in Brazil were also forced to produce digital exchanges, and they soon complied. While India struck a deal with Alcatel only in 1982 and the first ESS plant, set up at the uneconomic Mankapur site, commenced production only in 1986, CPqD had designed its own digital SPC exchanges—the Tropico series—

using the latest TDS technology by the early 1980s, the 1,000-line Tropico R being transferred to private industry for production in 1982. The CPqD program was carried out in cooperation with universities and private industry, and the whole telecom digitalization program was supported by TELEBRAS' autofinanced telecom expansion from 1977 on.

The Korean ESS project started only in 1981, but by 1984 it had developed the basic module, the TDX–1, and had, by 1986, installed the 10,000-line TDX–1A, a size of system not developed indigenously in India until the early 1990s. In India, such R&D and expansion investments were much smaller, expansion being constrained by the switching bottleneck itself. The fact that India was both burdened with the inappropriate Pentaconta crossbar technology that could not reliably cope with its low-telephone-density and high-usage-per-line pattern, and advantaged by its telecom industrial, R&D and institutional base, should have impelled a much earlier digitalization decision. Subsequently, when the decision to set up CDOT on a CPqD-like technology-but-not-production center basis was made in 1984, after a three-year delay, the project was extremely underfinanced not only by MNC but even by CPqD and ETRI standards.

Liberalization in the TV Industry

In the TV industry, the key factors were wholly political at the overall policy level and were bitterly resisted in the DOE. It was the Minister for Information and Broadcasting, Vasant Sathe, who set the ball rolling. Sathe personally forced the decision in 1982 to introduce and massively expand color broadcasting and to import color-TV kits for assembly by local SSI firms, to be on the market before the Asian Games in November 1982. He was also behind the regularization of this supposedly one-time and temporary development, triggering a TV manufacturing boom based on assembly of imported kits from Japan, West Germany, and South Korea. In this he was fully supported, on the one hand, by Prime Minister Indira Gandhi, and, on the other, by the ITMA lobby. The decision to expand the TV broadcasting network nationwide so as to increase coverage from 17% of the population in 1981 to 70% by 1984 by installing 170 low-power transmitters was entirely politically motivated. It was designed to allow the government-controlled TV network to support Mrs. Gandhi's drifting government in its attempt to propagate an energetic image and win support before the general elections due by December 1984.

The rapid expansion of the network was made possible by the Indian national satellite INSAT 1-B becoming operational in August 1983. This made the expansion of TV broadcasting far cheaper and more efficient than a terrestrial network would have been. The expansion of broadcast coverage also multiplied the size of the potential market several-fold, leading in 1982 to the ITMA lobby, their appetites whetted by the Asian Games bonanza, launching a campaign in anticipation to allow the assembly of imported kit and to keep out large private and foreign firms.

It is at the level of interests both of the political leadership and of the ITMA lobby that the new color-TV manufacturing policy and the unprecedented expansion of coverage must be explained. This policy was strongly resisted in the DOE, especially by those supporting indigenous color-TV development efforts by ECIL, those who considered it a nonpriority item, and those who wanted the components industry to develop, for which television was the largest source of demand. That demand would not materialize if manufacture was to be assembly of imported kit. Kit imports would also hinder standardization around specific components to help high-volume production of selected components such as linear ICs and discrete semiconductors. In the event, these forces were defeated.

INEFFICIENT ISI IN THE 1980s DESPITE RECOGNITION OF THE NEED FOR COMPETITIVENESS

Liberalization in the 1980s through to 1991 was not planned in accordance with any overall electronics strategy coordinating intersegmental development, despite the existence of a dedicated policy-planning and implementation agency for precisely the purpose of integrated development of the electronics industry. Color-television, computer, and component policies had independent origins and causes. These different strands were pulled together in March 1985 in the Integrated Policy Measures in Electronics, in tandem with the liberalization of the overall industrial policy regime.

The only exception to this pattern appears to be the New Software Policy of December 1986. It was a logical sequel to the computer boom following the 1984 computer policy and was deliberately delayed.[70] The New Software Policy allowed free import of software for stock and sale at 60% duty, and hardware for software exports.[71] The aim was to promote domestic software development for export on the "flood-in/flood-out" principle enunciated by Seshagiri, rejecting ISI as a futile enterprise in the extremely fast-changing software field. India was not to try to reinvent the wheel, but to use imported hardware, systems software, and software tools to develop custom application software for export to foreign clients for their specific needs.

But even this liberalization of electronics policy in the 1980s did not bring about selectivity and adequacy of investments, despite the recognition in principle of the imperative of competitiveness. Basically, it did not break out of the 1970s pattern of fragmented and subcritical investments in too many import-substituting production and R&D ventures without adequate demand for optimal scale created by appropriate complementary investments. It should have been clear that the critical importance of economies of scale at the production run level implied the need for selectivity and standardization in the choice of equipment and components. Economies of scale at the firm level were also critically important in both the components and most equipment segments, especially consumer goods. Discounts on bulk purchase of inputs and large revenues necessary for lumpy and risky investments in expansion and R&D were indispensable to competitiveness.

Furthermore, firm size enhances negotiating power in the acquisition of technology. For an ISI strategy in a developing country to have had a chance of success in creating competitiveness required massive complementary demand-generating investments. Selectivity would have to be exercised in choice of products for entry into, and standardization of products necessary for scale economies at the equipment level and for standardization and scale at the components level. The question that forces itself on the observer of Indian policy in comparison with that of Brazil is: why did India, despite the existence of a specially created EC and DOE since 1971, fail to implement a planned, coordinated, and phased policy liberalization?

And, as in the 1970s, the liberalization of the 1980s was accompanied by intensified ISI in high-tech equipment and components. What were these, and why were they undertaken? One strand in continuing advanced ISI was in aerospace/defense and consisted of projects such as the Army Radio Engineering Network, indigenous radar development efforts, and tank and microwave electronics. Other than these, there were strong indigenization thrusts in key equipment and components led by the public sector. ECIL signed a contract with Control Data Corporation in 1986 to manufacture its Cyber 180/810 and 830 series of mainframe computers at Hyderabad. The machines were essentially obsolete by 1986 and would certainly be so by the time they became commercially available and indigenized.

Why was such an effort undertaken after the TDC series failed to keep up with changing technology, despite total protection in an age when technology was not changing as rapidly as it had in the mid-1980s? According to former and then current top executives of ECIL, the Cyber series could not be competitive. The motive behind the decision to manufacture these was clearly political. The government opted for a modicum of self-sufficiency in large systems, even if those systems were technologically less than state-of-the-art. The CDC deal cost $9.5 million lump sum plus 5% royalty. The then managing director of ECIL conceded that the motivation for the contract was that the TDC series had become obsolete, and a large-system capability as contemporary as possible had become necessary. He also conceded that it would be near-impossible to indigenize the Cyber series with commercial success in an era of transition to minisupercomputers. CDC was willing to transfer technology but were not interested in equity participation—a sign of obsolescence.[72]

In telecom, a technology transfer deal was signed with CIT-Alcatel of France in 1982 for the manufacture of its E–10-B digital main exchanges by ITI at Mankapur. The first 500,000 line factory started production in August 1986. There were then three collaborations each for EPABX and for trunk exchanges. CDOT was set up in August 1984, with the mission of developing an indigenously designed family of exchanges ranging from the small (128-port) RAX through the 128-line EPABX to 16,000-port MAX. CDOT's chief, Sam Pitroda, was given Rs 360 million and 36 months to develop an indigenous family of exchanges. These were to be based as far as possible on indigenously designed components. ICs for the E–10-B were also to be locally made on a priority basis, such that 60% were to be local from 1989 on. The CDOT idea was originally canvassed by Pitroda from 1981 on, on visits to India. It was rejected initially, but Pitroda's persistence paid off, and a cabinet decision of February 1984 set up CDOT. The motive was ISI for both foreign exchange saving and technology development in communications equipment at a time when there was a political imperative to become more self-sufficient in switching equipment. In addition, Pitroda's exceptional technological record and his promise of economic efficiency and indigenous component development fit in with the self-reliance objective.

In components, there was a vigorous push toward indigenous capability in IC design and manufacture. A National Microelectronics Council was set up in January 1985 to formulate policy for the development of LSI/VLSI ICs.[73] SCL moved into production in April 1984 in 5-micron-linewidth chips based on American Microsystems technology. It developed 3-micron technology on its own and intended to move to finer linewidths, reaching 1.25 micron by 1990. The focus was on CMOS technology. Its collaborators were AMI-Gould in telephone pulse dialers, Rockwell in microprocessors for its microcomputer module, and Hitachi in digital clock modules. SCL was producing and exporting ICs for digital watches and clocks before it burnt down in 1989.

But the main thrust in microelectronics policy and in the reconstruction plans for SCL was to acquire the capability to design and manufacture ASICs for specialized applications in defense, space, and telecommunications. SCL had tied up with CIT-Alcatel for technology transfer for custom LSI devices for the E–10-B and codecs for PCM equipment. The viability of SCL depended heavily on coordination between ITI and private companies manufacturing telecom peripherals, including EPABXs. It also required the DOE to coordinate IC production requirements with SCL's program. There has been criticism that the telecom collaborations were not coordinated at the components indigenization end with SCL and BEL, and that three collaborations each in EPABX and digital telephones in the mid-1980s adversely affected IC standardization. But more than one collaborator may have been necessary to prevent any supplier gaining monopoly power.[74]

Despite this thrust, the microelectronics program and SCL have received very little investment by world standards, and that, too, was fragmented. The total investment over 1985–90 was projected to be Rs 4150 million for both production and R&D in microelectronics. SCL has received outlays of Rs 96 million (1985–86), Rs 150 million (1986–87), Rs 100 million (1987–88) and Rs 40 million (1988–89); the remaining money was spread over a large number of institutions and projects—and this in a world context where $200 million and $500 million were estimated in 1986 to be the Minimum Economic Scales at plant and firm levels, respectively, in ICs, and where Korea was investing $1 billion for ICs alone during 1985–90, highly concentrated in its largest firms, Samsung, Gold Star, KEC, and Hyundai. The price of a large-scale CMOS technology chipmaking plant in the late 1980s exceeded $100 million.[75] Thus even the special-thrust microelectronics segment shows that familiar faults of the 1970s were being repeated, despite awareness that there are critical masses of investment. As V. Mohan, Managing Director of SCL, put it, "Government has taken a view that for SCL, commercial considerations will have to be subordinated to the basic objective of technology development to meet the strategic needs of the country, in particular that of the Department of Space, Ministry of Defence and DOT."[76] This expresses in a nutshell the priorities of Indian microelectronics policy.

The Geopolitical Background to the Self-Sufficiency Drive in High Technology in the 1980s

Furthermore, from 1987 on, there was a renewed turn toward ISI, again motivated by strategic self-reliance and economic considerations. A DOE policy paper of December 1987 for the prime minister recommended encouragement of indigenous technology but also suggested using market access as a bargaining chip for transfer of technology from MNCs.[77] It recommended the import only of basic technologies, modification and upgrading to be done by local R&D, and it emphasized the importance of indigenous technology among liberalizing statements. Almost all of the DOE's projects and institutions—SCL, CDOT, CDAC, CDOME, NRC, and TDC projects, SAMEER—as well as BEL and ECIL have been concentrating on self-reliance at all costs. Not one initiative, with the exception of software, has been concerned with international commercial success.

This can be explained only at the level of "high political" national security considerations. This level of policy influenced electronics policy directly, although the policymakers at the electronics level and, in the public sector, BEL, ECIL, and SCL were equally aware of such overarching considerations. The era of gradual liberalization of electronics policy and the shift in the role of the state from direct producer to infrastructure provider and facilitator since 1981 has also been one of the worsening of India's strategic environment and the possibility of being subjected to technological sanctions.

India's regional security environment deteriorated in a major way after the Soviet invasion of Afghanistan in December 1979. Pakistan once again became a close ally of the United States as a vital arms supply conduit to the Afghan rebels. It received a massive US arms aid package amounting to $3.2 billion during 1982–87 and was slated to get $4.02 billion during 1988–93—a program that was terminated in 1990. Pakistan–China cooperation was also stepped up. Most ominously for India, Pakistan continued its clandestine nuclear weapons program, with the United States turning a blind eye. These developments implied that India was under pressure to match Pakistan's conventional and clandestine nuclear buildups. The implication for India was that it would have to have the financial and technological wherewithal for a conventional buildup and keep open a nuclear weapons option. But both these imperatives would require access to high technology or indigenous development of the same, especially in large computers and telecom systems with a "total system" integrated data-handling capability for air defense and command, control, and communication capabilities. This, in turn, implied access to the latest equipment and ASICs for defense and space applications, or indigenous technological capability in the same. Any overt move toward nuclearization would certainly lead to the tightening of technology transfer restrictions by the United States and other major powers, including the Soviet Union.

The Reagan Administration instituted unprecedentedly strict controls on the transfer of technology to the Soviet Union and as part of a revived attempt to strengthen export controls on dual-purpose and other sensitive high technologies and tightened export controls in general, including to Third World countries from 1981 on. The main purpose was to blockade the Soviet Union technologically. However, the policy also had purely foreign-policy motivations. India was perceived by the Pentagon to be close to the Soviet Union, the latter being its main arms supplier. India came to be included in 1983 among the countries suspected of being a possible conduit—witting or unwitting—for the smuggling of restricted high technology to the Soviet Union.[78]

The Export Administration Act of 1979, extended in March 1984, controlled exports of high-technology items not limited to the strictly military—those on the munitions list requiring clearance from the Office of Munitions Control in the State Department and those on the COCOM Commodities Control List, which includes dual-use and other sensitive technologies. It controlled a wide range of advanced technologies that could possibly have strategic potential. In April 1984 President Reagan approved a major expansion of the Pentagon's role in export clearances given to *"sensitive" but not classified goods to noncommunist countries*. In an interagency compromise the Pentagon agreed to restrict its country list to any 15 "risky" countries at a time. The United States also began to apply "extraterritorial" controls by threatening European allies and neutrals with loss of access to US technology unless they tightened their controls. Under the new dispensation, the Commerce Department and the Pentagon jointly were to review license applications from exporters in six technologically sensitive categories: (1) electronic and semiconductor manufacturing equipment; (2) test and measuring equipment; (3) microcircuits and ICs; (4) computers; (5) silicon and other components; (6) sapphire substrates used to make microchips.[79]

India had already been on the list of countries to which export licenses for high-tech equipment were frequently delayed or denied because of its 1974 nuclear test and its refusal to sign the Nonproliferation Treaty. From the early 1980s on, various high-technology products were denied to India. The United States declined permission to Sweden to proceed with the sale and coproduction in India of the Viggen fighter, since it contained key US-made components. Other items denied included a DEC VAX 11/780 computer for BARC.[80] US policy toward Indian high-tech purchases was not one of flat rejection, but one of prolonged delay—sometimes as long as two years—in clearance. These developments created great anxiety in the Indian defense, aerospace and electronics policy communities.

This was the background to the Indo–US Memorandum of Understanding on High Technology of 1984–85, negotiated and signed by India as it had become perfectly clear that it was the necessary condition for any transfer of advanced technologies. The first part of the MOU, outlining the basic principles, was signed in October 1984. The second part, outlining the procedures for implementation, was negotiated by March 1985 and signed in May 1985. The MOU, however, was not a sufficient condition for, and did not guarantee clearances for, technology transfer; it only made them possible—specifically, in three areas where licenses were being subject to prolonged delays: (1) LSI/VLSI ICs and plant and equipment for their manufacture; (2) telecom equipment; (3) "certain other areas" no doubt including computers.[81] Under the MOU, the United States demanded inspection rights for sensitive equipment transferred—above all, a Cray XMP–14, which was eventually transferred in November 1988. The MOU cleared the road for a technology transfer boom. This has not been an automatic process, and some licenses are still denied. Part of the Indian motivation for the MOU was access to defense electronics.[82]

The key point in all this is that Indian electronics policy in the 1980s had to keep this political background in mind and continue with indigenization-at-all-costs projects for fear of loss of access to high technology. There is clear evidence that the political climate of actual or threatened restrictions on technology transfer by the Reagan Administration was a crucial determinant of the continuing ISI effort in high technologies.

In components and materials, the Task Force on LSI/VLSI clearly states that "A local microelectronics base, particularly with regard to custom and semicustom circuits, is essential for future indigenous development of defense equipment within the country."[83] The industrial electronics segment, which is LSI/VLSI-intensive, is almost totally import-dependent. In an environment of threatened controls on ICs, there was a felt need to produce them locally, particularly custom chips. As the chairman and managing director of SCL, V. Mohan, put it while referring also to capital equipment and raw materials: "Our microelectronics program can be made independent of the export regulations of the vendor countries and can be made truly regenerative only through indigenous availability of equipment and materials."[84]

The clearance for the American Microcircuits-SCL microprocessor deal took months for what were by then obsolete (5-micron-linewidth) chips. The experience must have influenced policymakers, particularly as the negotiations were conducted by a senior DOE official who was well known to be a proponent of self-reliance in key technologies. Motorola refused to sell 16-bit microprocessor technology to SCL in 1986. The National Silicon Facility's deal with Hemlock Semiconductors for hyper-pure silicon manufacture (the basic semiconductor raw material) was held up. It was felt absolutely necessary to acquire LSI/VLSI capability for defense and aerospace ASICs because India was being subjected to restrictions on ASICs, especially radiation-hardened devices for potential defense and space applications.[85]

In computers, clearance for Control Data's Cyber 170/730 system eventually came through only in 1984, after an Indian assurance that it would not be used to service the DAE. The Control Data Cyber 180/810 and 830 came through in 1986 only after prolonged clearance delays. According to the then managing director of ECIL, these delays were a way of keeping Indian policymakers on tenterhooks, hampering their planning.[86] According to another former senior DOE official, the policy of delays in export licensing since the mid-1970s was meant to derail Indian projects. Delays were undeclared sanctions, he said, because they effectively halted investments in a program until the product became obsolete or would be so by the time it materialized after the clearance was obtained.[87] It was feared that the ultimate objective was to induce India to sign the Nonproliferation Treaty—something that Indian policymakers and political opinion across the spectrum were not willing to do.

To sum up, the principal reason behind India's apparently contradictory continuation of an ISI thrust in advanced electronics in the 1980s despite recognition of the need for a liberalized, efficiency-oriented ISI with a redefined role for the state, was political and strategic—the threat of cutoffs of technology for political reasons hanging over their heads. Indian policy might have been far more strictly economically driven, had the regional and international geopolitical environment been more relaxed.

STRATEGIC CAPACITY, POLICY FLEXIBILITY, AND THE POST-1991 SHIFT

First, it is important to note that the new strategy had its origins in the DOE from late 1990 on, after N. Vittal took over as Secretary, DOE, and a year before the launching of stabilization and structural adjustment in July 1991 under the new Congress government.[88] This was essentially because of fresh thinking at the top in electronics policy. It was realized that with a 31% import intensity of production, the industry would have to be restructured to become competitive. The industry was conceptually segmented into areas where applications were more important and areas in which it could become a competitive manufacturer. Three policy papers were prepared from September 1990 in consultation with industry, and a consensus was created by March 1991 on the need for freer retention of foreign exchange earned by exporters and freer input imports, if industry was to become efficient. However, nothing moved until the fortuitous July 1991 changes at the overall policy level, when a foreign exchange regime of this type came into being. The seachange in the state elite's strategic priorities and economic grand strategy both removed ideological constraints as well as reinforced the pressures to change electronics strategy by putting a premium on exports, and liberalizing imports and FDI.

However, it took one year from its conception in August 1991, in a DOE brainstorming session, to make the EHTP scheme official policy. For this, Vittal had to get past objections from the Ministries of Commerce and Industry and the Customs authorities of the Finance Ministry. However, with the new trade policy removing import licensing (quantitative restrictions) for inputs and making the rupee convertible on trade account by 1993, the pressure to export to access inputs for domestic hardware firms diminished. Therefore the EHTP scheme, which was designed precisely to turn Indian hardware firms into contract manufacturing exporters in alliance with MNCs, has not been as successful as hoped. Nor has it attracted FDI in a big way. The Commerce ministry, which runs the existing EPZ/EOU schemes, was also uncooperative in promoting the EHTP scheme for electronics alone.

As for the other major initiative—that of promoting software exports—the DOE, again at Vittal's initiative, sponsored over 90 64-Kbps-bandwidth international datacom channels during 1992–94 while whittling down the resistance and rates of DOT/VSNL, including bringing pressure on the latter through the Committee of Secretaries (administrative heads of government departments). The telecom bureaucracy and unions of DOT have traditionally resisted all attempts at reform, separation of functions or even technological modernization. It is interesting to note that all the changes in telecom policy originated from pressures from the overall policy level either because that level was affected by the aggregate effect of inefficiency affecting the interests of vast swathes of the economy, or because of changes in thinking induced by influential individuals.[89]

The DOE's initiatives, since 1992, in transforming the role of the private sector through the industry associations toward facilitative measures for export orientation, like the formation of consortia for international marketing and so forth to lower costs, have also not met with success. All the associations—MAIT, CETMA, ELCINA, Telecom Equipment Manufacturers Association (TEMA), and even the National Association of Software and Service Companies (NASSCOM)—have taken positions typical of long-protected home-market-oriented manufacturers, asking for easier and cheaper input imports, fiscal and financial concessions, and so forth, but they have resisted consortium approaches. The key issues have been those of whose product and whose component supplier should be the industry standard. There has been no reorientation toward exports on the part of the hardware industry, and most foreign collaborations have been undertaken in order to buttress positions in the domestic market. Thus the attempt to use peak private sector organizations as agents of policy implementation, *chaebol*-style, has not worked, due to the path-created weak, fragmented, and uncompetitive character of the private sector. In fact, FDI in home market-oriented joint ventures has accentuated the Indian partner's dependence on the foreign partner for technology and inputs, including loss of control in some cases such as HCL-Hewlett Packard and Tata-IBM.

Lastly, as for the rebuilding of SCL for an independent ASIC capability, its wafer-fabrication plant, which burned down in 1989, has not yet been rebuilt. The constraints were interagency coordination and at the overall policy level. Cabinet clearance for the project took inordinately long. Due to cost underestimates by the DOE, a second cabinet clearance was obtained in 1994, subject to support by strategic users such as Defense and the DAE. The Planning Commission asked the DOE to raise funds from the users, who were unable or unwilling to pay. The project has been on hold since November 1994.

The political economy of the post-1991 policy shifts throws light on several facets of strategic capacity and policy flexibility. What was attempted was a shift in the role of the state from producer and regulator to facilitator to implement a redefined strategy that focused on competitiveness. In this the initiatives came from the electronics policy level, with fortuitous facilitative changes occurring in the overall economic–ideological, industrial, and trade policy regimes later. This happened primarily because of fresh thinking at the top in the electronics policy under the tenure of N. Vittal as Secretary, DOE, from June 1990 to July 1993. The DOE remained uncaptured by vested interests. However, these attempted policy shifts have been only partially successful. The constraints have been those of poor interagency coordination, despite the overall industrial policy change, especially resistance from the DOT, and a weak private sector that represents an accumulation of path-created vested interests, not amenable to being used as a vehicle to implement the new strategy. However, path-created ideological legacies as a constraint on policy flexibility disappeared with the overall economic liberalization, only underlining the extent to which they had been a constraint earlier.

CONCLUSION: POOR COORDINATION AND POLICY INFLEXIBILITY

The overall picture that emerges is not one of a Korean-style neatly phased shift toward liberalization and more advanced products, but one of uncoordinated developments, with independent origins at different policy levels, despite the existence of dedicated policy institutions for integrated development of the electronics industry. The causes of policy shifts were typically at both policy levels, the overall determining or constraining the electronics level, and due to a mixture of causal variables, ideology, interests within and outside the state (and public enterprise), and "high political" considerations of national security and/or leadership grand strategy and its strategic priorities. The main point that we wish to emphasize is that *policy flexibility* was extremely limited at both electronics and overall policy levels, and economic policy-institutional consolidation was absent at the overall policy level and therefore limited at the electronics policy level because of its dependence on the overall level, making rapid and appropriate policy shifts in response to changing international and domestic market and technological conditions very slow. And this was due not only to the constraint of interests created by the path taken but also crucially in most cases until 1991 by the constraint of ideology that underlay the nationalist–statist "socialist" ISI model, going against which would invite powerful political attack and delegitimization—opposition to MNCs, foreign technology, large private firms, liberalized imports, pro-public and small enterprise, and self-reliance defined in self-sufficiency rather than competitiveness terms.

NOTES

1. World Bank, "India"; India: Ministry of Industry: Bureau of Industrial Costs and Prices, "Report on Electronics" (December 1987).

2. World prices, including cost, insurance, and freight.

3. World Bank, "India," Vol. II, pp. 15–16, Tables 1.8–1.10.

4. World Bank, "India," Vol. II, p. 12.

5. Wavesoldering and flowsoldering machines automate the process of soldering components onto printed-circuit boards in electronic equipment, a crucial step in the manufacturing process, which was formerly labor-intensive.

6. Interviews with chief executive, Weston, and former ITMA President, 3 October 1987, New Delhi; and with DOE official, 27 March 1987.

7. Ashoka Mody, "Institutions and Dynamic Comparative Advantage" *Cambridge Journal of Economics*, 14 (1990): 291–314.

8. Interviews with Marketing Director, Philips, 22 January 1987, Bombay; and with Chairman and Managing Director, Asha Pavro (Pvt.) Ltd., 16 January 1987, Bombay.

9. For a restatement of dependency allowing for peripheral autonomy and development, including bargaining power with foreign capital in development policy, see Stephan Haggard and Tun-jen Cheng, "State and Foreign Capital in East Asian NICs"; Johnson, "Political Institutions"; Evans, "Class, State and Interdependence"; Deyo, "Coalitions, Institutions"; all in Deyo (ed.), *The Political Economy*; Evans, "State, Capital." For the concepts of ideological guerrillas and pragmatic antidependency ideology, respectively, as constructs making for policymakers' ideology as a prime explanatory variable in Brazilian computer and Argentinian nuclear policy, see Adler, "Ideological Guerrillas," and *The Power of Ideology*.

10. For the best-known writings on this subject, see Douglass C. North, *Structure and Change in Economic History* (New York: W. W. Norton, 1981); Mancur Olson, *The Rise and Decline of Nations: Economic Growth, Stagflation and Social Rigidities* (New Haven, CT: Yale University Press, 1982); Anne O. Krueger, "The Political Economy of the Rent Seeking Society," *American Economic Review*, 64, no. 3 (June 1974). For India, see Pranab Bardhan, *The Political Economy of Development in India* (Oxford: Basil Blackwell, 1984).

11. Evans, "Class, State"; Haggard and Cheng, "State and Foreign Capital."

12. Ashoka Mody, "Planning for Electronics Development," *Economic and Political Weekly*, 22, no. 26 (27 June 1987); World Bank; Bureau of Industrial Costs and Prices, "Report on Electronics."

13. Lall, "India," pp. 562–563.

14. Interview with a former key AEC and EC/DOE official (30 March 1987) suggested that a device test was possible as early as 1966 and that had Bhabha not died, the program would have been accelerated and a test conducted long before 1974. One can only speculate that political pressures against an early test in the late 1960s would have been at least partly based on the much greater vulnerability to cutoffs of equipment and technology at a time when local production of much of such equipment had not been established.

15. This was strongly suggested and explicitly stated in interviews (22 and 30 March 1987 and 21 October 1987, respectively) with two former key electronics policymakers earlier associated with the atomic energy establishment, including one referred to in Note 14. However, this was formally denied by the then EC chairman, M. G. K. Menon (interview, 14 April 1993), who denied any connection between the moves for electronics self-sufficiency and the nuclear program.

16. The following account of the economic and political crisis of the mid-to-late 1960s and the resulting shift in economic policies is indebted to the detailed and comprehensive discussion of it in Francine R. Frankel, *India's Political Economy: The Gradual Revolution 1947–1977* (Delhi: Oxford University Press, 1978), chs. 9 and 10, pp. 341–433.

17. In this section we draw heavily on the analysis in Frankel, *India's Political Economy*, chs. 9, 10, and 11, pp. 341–490.

18. Edquist and Jacobsson, "The Integrated Circuit Industries," pp. 30–32, 35, 53–54; Wade, *Governing the Market*, pp. 312–319.

19. Grieco, *Between Dependency and Autonomy*, ch. 5, pp. 103–149.

20. Interviews, with former key DOE official, 27 and 30 March 1987, New Delhi; with former EC member and public enterprise chief executive, 21 October 1987, Hyderabad; with former defense public enterprise chief executive, 20 October 1987, Bangalore.

21. Interviews with three former ECIL chief executives: 21 January 1987, Bombay; 21 October 1987, Hyderabad; 22 October 1987, Hyderabad, for material in this and the next two paragraphs.

22. Philips India was renamed PEICO after this, despite retaining management control; however, for recognizability we refer to the company as Philips throughout.

23. Interviews with Marketing Director, Philips, 22 January 1987, Bombay; and with Chairman and Managing Director, Asha Pavro (Pvt.) Ltd., a subcontractor for radios, 16 January 1987, Bombay; Philips Annual Reports, various.

24. Interviews with Marketing Director, Philips, 22 January 1987, Bombay; and with former key DOE official, 30 March 1987.

25. According to Marketing Director, Philips, interview, 22 January 1987, Bombay.

26. Ashok V. Desai, "Indigenous and Foreign Determinants of Technological Change in Indian Industry," *Economic and Political Weekly*, 20, nos. 46/47 (November 1985); Don Scott-Kemmis and Martin Bell, "Technological Dynamism and Technological Content of Collaboration: Are Indian Firms Missing Opportunities?" in ibid., pp. 1991–2004.

27. A point stressed by Ghayur Alam, "India's Technology Policy and its Influence on Technology Imports and Technology Development," *Economic and Political Weekly*, 20, no. 46/47 (November 1985): 2073–2080.

28. Interviews with former key DOE official, 30 March 1987, New Delhi; and with former public sector chief executive and key policymaker on components, 20 October 1987, Bangalore. From these sources it would appear that they did not consider the option workable for purely political reasons and hence made no effort to explore it seriously.

29. N. Seshagiri, "Econometric Study: Economical Scale of Production for Television Receiver Industry under Indian Conditions," *EIP*, 1, no. 1 (October 1973): 107.

30. Ibid. p. 114; "n" refers to the daily output of TV sets.

31. Interviews with Dyanora, 30 January 1987, Madras; and with chief executive of Weston, 3 October 1987, New Delhi.

32. Interviews with chief executive of Weston, 3 October 1987, New Delhi; and with DOE official, 27 March 1987, New Delhi.

33. Information from ITMA sources.

34. Vijay P. Bhatkar and N. Vijayaditya, "Manpower Study: Report of the Committee on Small and Medium Entrepreneurs: Implementation Strategy for Electronics and Communications," *EIP*, 2, no. 1 (October 1974): 67.

35. Ibid., p. 69.

36. Ibid., p. 70.

37. See all articles on SSI in *EIP* in the 1970s and the Sondhi Committee Report, Part I, *EIP*, 7, no. 6 (March 1980): 391.

38. Interviews with former key DOE official, 22 and 30 March 1987, New Delhi.

39. *ELCINA Directory*, 1984–85.

40. "Policy on Electronic Components," paragraph 6, in *EIP*, 15, no. 1 (October 1987): 40.

41. Interview with DOE official, 22 April 1991, New Delhi.

42. "Report of the Committee on Electronics Exports," *EIP*, 6, no. 2 (November 1978).

43. Interview with Mantosh Sondhi, 3 April 1987, New Delhi.

44. Interview with private sector components firm chief executive and former president of ELCINA, 8 December 1986, New Delhi.

45. Interview with private sector components firm chief executive and former president, ELCINA, 6 February 1987, Hyderabad.

46. Interview with executives, Dyanora, 30 January 1987, Madras.

47. Interview with ITMA President, 18 December 1986, New Delhi.

48. ITMA Press Releases, 29 May 1985 and 22 July 1985.

49. Interview with Marketing Director, Philips, 22 January 1987, Bombay.

50. Interviews with former key DOE official, 22 March 1987, New Delhi; and with Marketing Director, Philips, 22 January 1987, Bombay.

51. Grieco, *Between Dependency and Autonomy*, pp. 100–101; interview with former ECIL chief executive, 21 October 1987, Hyderabad.

52. Interview with former key DOE official, 30 March 1987, New Delhi.

53. Interview with former ECIL chief executive, 21 October 1987, Hyderabad; Grieco, *Between Dependency and Autonomy*, pp. 98–102, 128–147.

54. Interviews with named and other firms.

55. Interviews with key former policymakers.

56. Interviews with former Information and Broadcasting Minister Vasant Sathe, 20 August 1992, New Delhi; with former key DOE official, 30 March 1987;

with Ministry of Information and Broadcasting officials at various times, New Delhi.

57. World Bank, "India," Vol. II, p. 110; interviews with former key DOE and public sector officials. For example, the EC took two years in the 1970s to clear the on-line process-control system for the Bokaro Steel Plant.

58. World Bank, "India," Vol. II, p. 110.

59. World Bank, "India," Vol. II, p. 41.

60. Interview with former key DOE official, 30 March 1987, in which he generally blamed "the system" beyond the control of electronics policy.

61. Hans-Peter Brunner, "Building Technological Capacity: A Case Study of the Computer Industry in India, 1975–87," *World Development*, 19, no. 12 (December 1991): 1737–1751.

62. Grieco, *Between Dependency and Autonomy*, ch. 5, pp. 131–138.

63. Interviews with then incumbent and former key policymakers in the DOE for the following paragraphs. This account first appeared in E. Sridharan, "The Political Economy of Industrial Strategy for Competitiveness in the Third World: The Electronics Industry in Korea, Brazil and India" (Ph.D. dissertation, Department of Political Science, University of Pennsylvania, 1989), pp. 402–412, and is corroborated independently in C. R. Subramaniam, *India and the Computer* (New Delhi: Oxford University Press, 1992), pp. 43–77.

64. These changes of personnel were fortuitous.

65. It was Pitroda's lobbying that played a key role in the Sarin Committee deciding in favor of digitalization; interview with Sam Pitroda, 20 May 1995, New Delhi.

66. Telecom facts and figures from Brundenius and Goransson, "Technology Policies"; DOT Annual Reports.

67. For an account of the Indian cross-bar modification project and for technology development in electronic switching in the 1980s, see Sunil Mani, *Foreign Technology in Public Enterprises* (New Delhi: Oxford & IBH, 1992), pp. 67–118.

68. Interview, Sam Pitroda, 20 May 1995, New Delhi.

69. Interview with Sam Pitroda, 20 May 1995, New Delhi. Interestingly, in his first presentation on the viability of an indigenous digital ESS design effort before the Prime Minister in 1981 he did *not* mention his association with the Brazilian effort since 1976 and its—by 1981—successful indigenous design of a small TDS electronic exchange. One can only speculate that had he done so, CDOT may have been born three years earlier, and the history of India's telecom and information technology industries may have turned out differently!

70. Interview with N. Seshagiri, Additional Secretary, DOE, 10 April 1987.

71. For a detailed discussion, see *Dataquest* (January 1987). For the policy text, see *EIP*, 15, no. 1 (October 1987): 50–55.

72. Interview, 21 January 1987, Bombay.

73. For the material in the following paragraphs, see SCL and DOE Annual Reports, "Report of the Task Force on LSI/VLSI Technology," Part 1A and Part 1B, *EIP*, 11, nos. 4/5 (January & February 1984); various issues of *EIP*.

74. SCL Chairman's Speech in SCL Annual Report 1985–86 and interview with former components policymaker and public sector chief executive, 20 October 1987, Bangalore.

75. Edquist and Jacobsson, "The Integrated Circuit Industries," p. 11. CMOS technology is used in computer memory chips.

76. Chairman's Speech, SCL Annual Report 1985–86, p. 2.

77. "Strategic Initiative for Electronics," DOE (December 1987).

78. Joseph Fitchett, *International Herald Tribune* (Paris), 2 June 1985.

79. *Electronics Business*, 1 June 1984.

80. *Telegraph* (Calcutta), 14 May 1983; Rohini Nilekani, "Plugging the Leaks," *Imprint* (Bombay), November 1984, p. 11; T. N. Ninan, "Restrictive Practice," *India Today* (New Delhi), 31 August 1984, pp. 100–101.

81. *Economic Times* (New Delhi), 15 May 1985.

82. Which was discussed with US Undersecretary of State for Defense Policy, Fred Ikle, in May 1985, *Hindu* (Madras), 4 May 1985.

83. "Report of the Task Force on LSI/VLSI Technology," Part 1A, *EIP*, 11, no. 4 (January 1984): 156.

84. SCL Chairman's Speech, SCL Annual Report 1985–86, p. 4.

85. Interviews with key former and then current DOE officials, March and October 1987; latter emphasized the fear of controls on defense-required radiation-hardened devices.

86. Interview, 21 January 1987, Bombay.

87. Interview with former key DOE official, 30 March 1987.

88. Information in this section derived from interviews with N. Vittal, Secretary, DOE, 18 December 1992, New Delhi; with Pronab Sen, Economic Adviser, DOE (1990–95), 16 December 1992 and 25 May 1995, New Delhi; with Sam Pitroda, 20 May 1995, New Delhi; also N. Vittal, "My Years with the Department of Electronics: Management of a Technical Department in Government," *EIP*, 20, no. 11 (August 1993), and press reports.

89. According to Sam Pitroda, interview, 20 May 1995, New Delhi, his equation with Rajiv Gandhi was the key factor in permitting private manufacture of small exchanges in 1984, the corporatization of MTNL and VSNL in 1986, the formation of the Telecom Commission in 1989, and, incidentally, the setting up of CDAC in 1988.

7

Conclusion: Reconceptualizing Strategic Capacity

The foregoing chapters have analyzed the factors behind the evolution of electronics policy in Korea, Brazil, and India. In each case, the ultimate determinant in the choice of strategies was political. In this chapter, we consider the political requirements for the success of alternative industrialization strategies for achieving international competitiveness in new industries. Our focus is on the Indian experience, but the discussion has broad implications for the comparative political economy of industrialization.

STRATEGIC CAPACITY: DOES IT EXPLAIN THE ELECTRONICS EXPERIENCE OF BRAZIL AND INDIA COMPARED TO KOREA?

The Concept of Strategic Capacity

It is now commonplace to argue that the East Asian NIC strategies were not quite as economically liberal as had been made out earlier, and that they were state-defined and state-managed or *strategy-led* models of industrialization. It is also accepted that their success was conditioned by a highly specific political context, domestic, regional, and global.

The concept of *strategic capacity* is crucial to an understanding of the diversity of experience of these three countries. We have outlined the notion of strategic capacity as defined by Deyo and extended in effect, though without direct reference to it, by the work of Wade and of Evans and Tigre, to explain strategy-led industrialization in the NICs. We now attempt to review the strategic capacity model and refine it to augment its explanatory power. Following that, we examine the conditions that made the Korean electronics strategy possible and then compare it with that of Brazil, to try to explain the differences. Our focus is on why the Brazilian state did not emphasize international competitiveness in its ISI strategy, despite the export imperative resulting from the debt crisis. We then compare the ISI strategies of Brazil and India—for Brazilian ISI was still far more efficient than India's in terms of narrowing the gap with world prices and technology, not to speak of overtaking India in computers and telecom, despite the latter's head start. Finally, we examine whether it was the lack of strategic capacity that explains the Indian experience, further refining the strategic capacity concept in the process.

We would argue that although the concept of strategic capacity is useful, it is incomplete in crucial respects. Firstly, while political closure may be necessary for ensuring that the state *can decide* what goals to set and what sort of strategy to follow, and economic institutional consolidation or coordination may be necessary to ensure that the state *can carry out* the strategy, what is left unclear is what sort of strategy the state *will decide* to adopt and implement. We argue that this is critically important to understand for extrapolating the possibilities of strategy-led industrialization, since states with the power to choose strategies will not necessarily choose similar ones and will not necessarily have the same objectives, such as international competitiveness, in mind. It is vitally important to be able to explain why this is the case. For while the creation of internationally competitive industries is vital to sustain economic growth and self-reliance even in an ISI strategy, this may not be the political priority of any particular country. But if it is not, then industrialization may not be sustainable in the long run. Rather, one may see successive rounds of ISI dependent on imported technology and capital goods, and a deepening debt crisis, as Evans argues has been the case in Latin America. Hence strategic *capacity* alone in the state-as-machine (implementation capacity) and state–society relations (autonomy) senses is inadequate; for assessing the prospects for successful strategy-led creation of new industries in the Third World, it is theoretically crucial to define the conditions under which a state with the autonomy to choose *will*, in fact, choose to make international competitiveness a goal for the new industry it plans to develop—among other objectives, if necessary. The question of strategic design—of state "leadership" or "followership," big or small—follows after this.

For this, one needs to understand the *strategic priorities* of the state elite. The concept of strategic capacity determined by the autonomy and implementation capacity of state institutions needs to be situated within the framework of the political–economic grand strategy of the state elite, in turn determined by their strategic priorities. This is not quite the same as Adler's emphasis on ideology, because the grand strategy of the state elite is a product of their own group interests, ideology, and the domestic and international political context, which, taken together, determine their strategic priorities. This enables us to get beyond defining the conditions under which we can predict whether states are able to decide and implement strategy, to explaining what they actually decide to choose and why.

Secondly, the strategic capacity concept leaves out the path-dependent dimension of policy flexibility as defined by Nayar. This needs to be further enlarged to include not only the constraints on flexibility imposed by interests created by past policy within and outside the state, but also ideology as a constraint on flexibility. For, as we explained in Chapter 1, the legacy of ideological positions taken in the past, when made central to political legitimacy, can be a significant constraint on policy shifts beyondcertain limits.

Strategic Capacity in the Korean Case

The paradigmatic case of such autonomy and implementation capacity, or political closure and economic institutional consolidation—later coordination and flexibility—would be the Korean state in the authoritarian period up to 1987. Its consolidated economic policy and implementation institutions would be the EPB and MTI (also the Blue House, MOST, MOC, and the NCA since the 1980s); its private sector counterpart would be the *chaebol*s.

Korea's three-stage export-led growth strategy in electronics fostered domestic firms capable of competing internationally in progressively more sophisticated products. We have already seen that the political basis for this state-orchestrated development trajectory consisted of the following: (1) a highly insulated state up to 1987, enjoying (a) a state-dominated partnership with domestic big business; (b) the exclusion of foreign capital from political influence, due to its absence to begin with and its entry regulated by the state from the outset; (c) the repression of all political opposition including, but not limited to, labor unions, the Left, and small business. (2) This highly autonomous state had the capacity to formulate and implement policy—strategic decision-making was centralized in the EPB, later coordinated with MTI (currently MOTIE—Ministry of Trade, Industry and Energy), the Blue House, MOC, MOST and the NCA, and the government controlled the banks. (3) It was dominated by a military elite, not immediately threatened, whose compelling political interest—and, hence, strategic priority—was to promote export-oriented growth for that was the only way to expand prosperity and secure legitimacy. (4) Strong private sector conglomerates that could undertake large, risky investments; (5) Policy flexibility, allowing the withdrawal of the state from closely directive guidance from the mid-1980s, due to the limited character of earlier intervention, which did not extend to state ownership or unconditional subsidization of industry. This, in turn, derived from the fact that ideologically the state elite was utterly pragmatic and result-oriented about growth. In Chalmers Johnson's terms, it was "market-conforming" not "market-repressing" in its interventionism, though, as we have emphasized earlier, it tried to anticipate and create comparative advantage for future competitiveness in the world market rather than channel investments according to existing market signals for short-term profits.[1] This characteristic and the regime's long-term holistic perspective was possible largely because of its political insulation.

Brazil: Lack of Strategic Capacity?

When we compare the Brazilian model with the Korean, we see strong apparent similarities in the nature of the political regime. The Brazilian military dictatorship of 1964–84 resembled Korea's in the following five important characteristics.

(1) Both were *authoritarian military regimes* that came to power after overthrowing elected civilian democratic or quasi-democratic governments.

(2) Both regimes *completely excluded the "popular sector"* of organized labor, the peasantry, the political and intellectual left, students, the middle class, and small business. The only groups outside the state apparatus that could possibly have any say in strategic decision-making were domestic and foreign capital. Thus, relations between the state elite and domestic and foreign capital, as well as the latter's interrelationship, became key factors defining the social–coalitional basis of strategic policy choices.

(3) While South Korea was an ally of the United States and in the US security orbit, hosting 40,000 US troops on its soil, Brazil was formally a nonaligned country. However, the similarity lies in the fact that Brazil considered itself a pro-Western state and was considered one by the United States, being staunchly anticommunist at home and in Latin America generally. *The implication was that Brazil could count on US support and faced no threat of cutoffs of high technology and no pressing need to import-substitute across the board* in high-technology industry and in defense production and R&D. That Brazil nevertheless did so with vigor, creating large armaments and capital goods industries, not to speak of computers and telecom, is another matter.

(4) In both countries electronics policy *originated within the state,* and not from the interests of private capital. In Korea, electronics policy was entirely state-defined, and it was so in a comprehensive manner as part of a general export-led growth and later "deepening" strategy, motivated by economic considerations. However, after about 1984, the state increasingly yielded its "big leadership" role for a complementarily supportive "big followership" role. In Brazil, it originated in different agencies of the state apparatus for computers and telecom, each independently, and was institutionally consolidated only after 1979 with the formation of SEI. However, it was motivated by nationalist–ideological and national security considerations, apart from economic ones. Economic considerations were defined differently. The concern was not so much with export earnings but more with creating opportunities for capital accumulation by nationally owned industry, especially in computers. However, by the early 1990s electronics policy was increasingly left to market forces as the National Model came under pressure.

(5) In both states *the banks were owned or controlled in major part by the state,* and credit allocation was used (in Korea) or potentially usable (in Brazil) as a instrument to implement electronics strategy. However, the functioning of such control was significantly different in Brazil.

These five similarities seem to imply a similar strategic capacity in Deyo's sense of the term. But we have seen that Brazil's electronics strategy differed considerably from Korea's and was considerably less successful, especially if measured in terms of international competitiveness. Why was this so? Were there any important differences that overrode the similarities noted above?

How, for example, was Korea able to exert control on the direction of domestic investment through controlling the banking system and channel it toward selected export-oriented industries, while Brazil was unable to do the same? How was Korea able to implement the strategy of picking winners while being able to withstand the political pressures of the losers? These question require an analysis of successive political regimes in Korea and Brazil since the early 1960s at a macropolitical level, in terms of their sociopolitical composition, strengths, and weaknesses.

The crucial political events and consequent regimes that need to be understood are the Korean coups of 1961 and 1971 (the 1979 coup did not change the type of regime so much as the personalities) and, in the case of Brazil, the coup of 1964. The regimes that emerged after the 1961 coup in Korea and the 1964 coup in Brazil abandoned the older phase of ISI and moved toward export-promotion and a new phase of ISI. The regime that emerged after the 1971 coup in Korea—the Yushin regime—was the one that promoted heavy and chemical industries from 1973 onwards, taking Korea into a new phase of its industrial development. As we have seen, these successive regimes in Korea and Brazil intervened in the economy in quite distinct ways, despite apparent similarities.

In Brazil, the post-1964 authoritarian regime was characterized by the institutional control of power, wielded by the military officer corps as a whole, unlike either the old-style *caudillismo* or the earlier populist-style authoritarianism of the Vargas era. The military took power "not as in the past to maintain a dictator in power (such as Vargas or Perón) but to reorganize the nation in accordance with the "national security" ideology of modern military doctrine."[2] The social base of the regime was a triple alliance of the military-dominated state apparatus, local capital, and foreign capital present in the economy as direct investment. Labor and the lower classes were excluded by direct repression until the late 1970s and early 1980s, when efforts were made to co-opt the unions through limited representation in a corporatist fashion. After suffering delegitimization in the devastating recession of the first half of the 1980s, the regime allowed elections in 1984, marking Brazil's return to democracy.

It is a crucially important fact that foreign capital was massively present in the economy from the inception of the military regime in 1964, and that it was in many ways in alliance with local capital rather than in conflict with it. Any hypothetical attempt to limit the operations of foreign capital meant politically taking on an entrenched presence allied to local capital in many of the most important existing industries, not to speak of endangering the health of MNC-dominated industries—and this in a situation in which the military regime had no base in the labor movement and needed the support of the United States and US-dominated international financial institutions. What is more, the Brazilian military was not a penetrative organization, with deep roots all over the country. Nor was there any external threat that it could use as a means to legitimate its rule. Therefore the only industries from which the state could possibly exclude foreign capital to protect domestically owned firms were new industries in which there was no existing foreign presence.

Not that the military regime had any such ideological bent. On the contrary, it was dominated by economists of orthodox persuasion representing the anti-ECLA, anti-ISI intellectual backlash. Fundamentally, the Brazilian military was weak along the dimension of autonomy, though it was strong in capacity, that is, control over economic resources at the level of the overall regime. It had to contend with a well-developed local capitalist class and a large foreign economic presence mainly allied to the former, neither of which it had historically dominated. For such a regime to intervene, even if it wanted to, in the economy in such a way as to force industry to develop along lines that could contradict their short-term market-determined interests in order to serve some long-term plan for competitiveness in regime-selected industries would have been very difficult politically. A regime strategy of picking winners from a field of relatively numerous and well-developed firms, many of them foreign and uninterested in helping Brazil compete in their home markets, was well-nigh impossible, even if desired. This is why, even though the state controlled the banking system as in Korea, its use of these powers was quite different from that of the Korean state as far as industrial policy was concerned.

Korea's constellation of social and political forces was markedly different. The country emerged from the Korean war with a very powerful, well-entrenched military regime that legitimated its rule by constant invocation of the communist threat from North Korea. The native capitalist class was relatively small and wholly subordinate to the state. Land reform had given the regime a base in the peasantry. Indeed, the National Agricultural Cooperation Federation, a deeply penetrative organization of cooperative farmers, was tightly integrated into the central bureaucracy. The regime was solidly supported by the United States, and the availability of US aid, in fact, gave it an extra degree of freedom from domestic capital. In the 1950s and 1960s, the Korean government did not rely so much on taxes, which reduced the need to negotiate with domestic firms or respond to their pressures. There was hardly any foreign direct investment in Korea, and hence no lobby of entrenched foreign firms to contend with.

After looking at the Brazilian state more closely, one can spot important differences that affected strategic capacity, despite the five similarities noted above. The key differences pertain to the Brazilian state's lack of autonomy from the two major components of the dominant class coalition in Brazil, foreign *and domestic* industrial capital. Unlike in Korea, the Brazilian state was far from being autonomous from these two dominant-class interests. As *dependencistas* have argued and empirically shown, the Brazilian industrialization model was one of triple alliances between the state, foreign capital, and domestic capital. Foreign capital, as Evans argues when comparing the Latin American to the East Asian NIC model, was an entrenched presence, allied in a multitude of ways with domestic capital, even while their interests clashed in other ways. This already-entrenched presence was the case in electronics across segments, severely limiting the autonomy of the state from both interests and constraining its nationalist initiatives. We have argued that this was a critical constraint in electronics, especially in the drive to establish national control in the telecom industry, in contrast to the new mini- and microcomputer industry, in which national control was established from the outset by market reserve. This was politically possible because of the absence of foreign firms.

The state lacked strategic capacity in Deyo's sense, since it could not achieve one of its preconditions—political closure. The state also lacked strategic capacity in that economic institutional consolidation was not achieved. Important segments of the banking and financial services sectors remained under private ownership and control and were usually part of domestic private industrial conglomerates. Institutional consolidation of electronics policy was delayed until 1979 and not really achieved until the National Informatics Policy of 1984. Even this consolidation was weak and constrained. Nationalist initiatives in computers and telecom that did arise within the state apparatus were greatly hampered in their implementation.

This is not to argue that the state was reducible to private interests. As we have argued for electronics and as Evans has pointed out, there was political space for the state to intervene massively in the sphere of production in aid of national—autonomist objectives. What is more important is the highly *constrained* nature of that intervention. Brazilian ISI strategy, unlike that of Korea, did not move rapidly toward achievement of international competitiveness. On the contrary, it was limited to domestic-market-driven development, with firms making investments according to criteria of profitability in the protected domestic market, reserved for nationally owned firms in computers. The consequences for the appropriateness of choice of product, scale, technology, and other factors relevant for international competitiveness guaranteed the inefficiency of Brazilian ISI. The domestic-market-driven character of Brazilian ISI was due to the state's inability to control credit allocation through the banks, and hence private investment and technology development in such a way as to anticipate and create comparative advantage *à la* Korea ignoring short-term efficiency considerations. The triple partnership was not unambiguously state-dominated as was the Korean model.

Thus, despite some broadly similar political conditions, the Brazilian state was unable to exclude from decision-making the above interests. State elites lacked the strategic capacity and, as important, the political motivation for an export-oriented efficient ISI strategy, as opposed to merely creating new industries and capturing for national capital the domestic market from imports or foreign firms in the economy. A Korean-style attempt to impose long-term priorities on Brazilian industry would have meant confrontation with the immediate interests of both national and foreign firms. Such confrontations would have seriously undermined and internally divided the state elite, given its penetration by its social—coalitional base, and endangered the regime's stability. Hence one can conclude that there was also no political motivation to follow such a course, quite apart from the question of strategic capacity.

India: Lack of Strategic Capacity?

The Indian case poses the question of strategic priorities of the state elite and path-determined ideological constraints on policy flexibility even more sharply. If we compare Indian with Brazilian ISI, we see that although national—autonomist objectives were far more assertively defined in India, the result was an even more inefficient and internationally uncompetitive brand of ISI. We have examined the course of and causal factors behind Indian electronics policy over the period. We now step back and try to answer the question of why India, despite a head start in planning, fared the worst in electronics.

Did India lack strategic capacity? Is this what explains the even greater uncompetitiveness and lack of attention to efficiency considerations than was apparent in Brazil? This hardly seems likely. Whereas in Brazil political closure was incomplete, in India the state excluded foreign capital from the dominant coalition. At the level of electronics policy, it also excluded domestic big capital, except possibly in computers after the mid-1980s. It also excluded "popular-sector" interests, except for small-scale industry. Even the latter did not really penetrate the EC/DOE. At that level, closure was essentially achieved.

However, at the level of overall industrial policy, there was no such autonomy from "popular-sector" pressures. India's parliamentary democratic framework allowed labor unions to function freely and resulted in legal protection of workers' rights, unlike in Korea and Brazil. Politically, it was extremely difficult to lay off surplus manpower or close down chronically unprofitable plants. Therefore, the East Asian type of export orientation based initially at least partly on labor repression in EPZs was simply not possible politically. In addition, the all-India character of the Congress party and the need to manage regional imbalances exerted pressures for a general backward-area-biased location policy, which was applied even in a quite inappropriate industry such as electronics, hampering its efficiency.

Moreover, in the late 1960s and early 1970s, the political and intellectual left was influential enough to influence overall industrial policy in a not just a pro-public sector—which by itself need not necessarily have been damaging—but populist "anti-big" houses and pro-small-business direction. It was also militantly and dogmatically opposed to the entry of foreign capital and technology. The possibility was not considered that foreign firms could be harnessed to the national effort in ways that did not threaten loss of sovereignty or the dominance of national capital, as was done in Korea. Given that during the 1970s the Indira Gandhi government campaigned on a populist platform claiming to be socialist and anti-imperialist, pragmatic policies toward foreign firms would have been politically damaging and ruled out for political reasons alone, creating historical path-determined ideological constraints on policy flexibility at both overall and electronics levels. *Policy flexibility*, therefore, was extremely limited at both electronics and overall policy levels, making rapid and appropriate policy shifts in response to changing international and domestic market and technological conditions very difficult. And this was due not only to the constraint of interests created by the path taken, but also crucially in most cases by the constraint of ideology that underlay the nationalist–statist "socialist" ISI model, going against which would invite powerful political attack and delegitimization—opposition to MNCs, foreign technology, large private firms, liberalized imports, pro-public and small enterprise, self-reliance defined in self-sufficiency terms.

The relative lack of political closure at the overall policy level meant that the state elite had to contend with diverse and conflicting interests. Far from being insulated, state elites were heavily penetrated by interests within the dominant class coalition and counterpressured by the Left since the late 1960s and early 1970s. The lack of autonomy at the overall policy level had its effect on the much more insulated electronics policy level, since the latter had to function within the former's constraints. Indian electronics policy faced a lesser problem of closure when compared to Brazil. However, at the level of the overall policy regime it, too, faced problems of closure, not only from big business with a vested interest in the protectionist ISI model, but also from "popular-sector" interests.

As regards economic institutional consolidation, this went much further in India than in Brazil at the electronics policy level, and it did so from much earlier on—that is, from the formation of the EC/DOE in 1970–71 reporting directly to the Prime Minister. Despite this organizational head start, Indian policy shaped up very slowly and incoherently to the windows of opportunity opened up by new developments in technology and market structures. This is explicable in terms of the constraints that Indian electronics policy institutions were subjected to from the overall policy regime. And at that level economic institutional consolidation did not take place. There was nothing at the overall policy-institutional level analogous to the EPB in Korea in the 1970s.

On the contrary, there were several different agencies that granted the various clearances necessary, including industrial licenses, import licenses, approvals for foreign collaboration, sanctions for foreign exchange, and long-term industrial credit. These bodies did not coordinate their activities. Indeed, there was no unifying legislation on such inextricably interrelated and interactive matters as foreign equity collaboration, terms of technology import, industrial credit, and capital goods import. Even in the matter of industrial licensing, enterprises had to run the gauntlet of MRTP and FERA, apart from the routine letter of intent and industrial license. The pieces of legislation governing each of these matters were separate and uncoordinated, and they were basically regulatory rather than promotional in approach and effect, even though each of these decisions was intimately interdependent.

For example, in importing technology and capital goods for a new project, one has to take into account the possibility that closely held high technology may not be available without allowing equity participation. Therefore, technology licensing rules on royalties and contract duration on the one hand and FERA on the other would need mutual adjustment or coordination; or, alternatively, there may be economies of scale involved that would necessitate licensing large houses if the project is to be viable, notwithstanding the restrictions built into the licensing policy, SSI preference, or whatever MRTP and FERA provisions may say. Industrial credit decisions by the public sector industrial financing institutions would have to coordinate with the above rules, whatever the existing credit priorities, in the larger interest of project viability.

Yet none of the key pieces of industrial policy legislation was drawn up with the goal of meshing with the others to promote efficient and competitive industry. Rather than have laws so worded that they merely laid down the general *principle* of promoting national ownership and/or control, leaving the definition and implementation of the principle to an integrated policy agency or coordinated interagency committees on a case-by-case basis depending upon the industry, technology, firms, and relative national bargaining power as in Korean and Brazilian laws, FERA, for example, laid down a specific 40% criterion for foreign firms across the board, regardless whether the law affected a computer firm or a soft-drink firm. Similar across-the-board laws were adopted on royalty, lump-sum fees, and contract duration limits in the matter of technology import, despite the extreme sensitivity of these matters to bargaining power and its interrelation with equity participation. Each law, whether industrial licensing, MRTP, FERA, small-scale reservation, technological collaboration, industrial credit, import licensing, or foreign exchange clearance, stood on its own.

Despite much talk of single-window clearance, this had been achieved only to a limited extent before the changes of the post-1991 reforms by the constitution of the Inter-Ministerial Standing Committee (IMSC) in the DOE in 1985 (for computers). And the members of this committee came to the table bringing their own independent sets of rules governing the interdependent decisions that they were to discuss. This lack of economic institutional consolidation or coordination at the level of the laws and agencies governing the overall industrial policy regime severely constrained the electronics policy regime, despite its early consolidation. The contrast with the encompassing powers of the Korean EPB and, in the 1980s, with the close interagency coordination between the Blue House, MTI, MOC, MOST, and NCA across all the above issues critical to investment decisions could not be greater. And it is the investment decision that is the critical one, as it determines the product, technology, scale, and organization, with consequences for future performance.

Clearly, then, the partial character of political closure and the lack of economic institutional consolidation or close coordination at the overall level hampered India's strategic capacity, in Deyo's sense. But is this alone an adequate explanation of Indian policy failures? Despite the lack of strategic capacity in a pure sense, it is possible to argue strongly for the Indian state's strategic capacity in segments of electronics, as Grieco does for computers. Grieco concludes that India was successful in improving the terms of its relationship with the international computer industry since the creation of the electronics policy institutions. From this, he concludes that the Marxist–dependencia school of thought on developing-country–MNC relations, which argues for basically unalterable structural dependence, is analytically inferior to the post-dependency bargaining school, which holds that developing countries can improve the terms of the relationship by policy innovations. His "key finding remains that Indian actors—the central government and enterprises not under its direct control—were able to undertake actions, which led to an improvement in India's computer industry according to objectives established in the mid-1960s."[3]

Our analysis supports his conclusion in other segments of the electronics industry, at least to some extent. From the point of view of creating an indigenous technological base and manufacturing capacity, one can even speak of evidence of strategic capacity. It is buttressed by the fact that despite the inadequacy of strategic capacity in a pure sense, the Indian state has been able to execute strategic industrialization and technology development programs even more complex than in electronics—for example, in nuclear, space, and defense industries—since the 1960s. This suggests that notwithstanding the weaknesses of strategic capacity due to incomplete political closure and economic institutional consolidation at the overall policy regime level, there were other factors at work in its failure to create an internationally competitive electronics industry.

While India may have fared poorly compared to Korea and Brazil—especially the former, and particularly with regard to international competitiveness—it certainly increased its domestic capabilities and for the time being its technological autonomy to a state "between dependency and autonomy," to use Grieco's phrase. Obviously, India did have some degree of strategic capacity, despite the constraints at the overall policy regime level. The question remains why it was not able to do much more with the strategic capacity that it did have, especially when combined with its institutional head start in electronics policy?

GETTING BEYOND THE STRATEGIC CAPACITY MODEL FROM THE INDIAN ELECTRONICS CASE

A useful starting point for getting beyond strategic capacity as a political economic explanation of industrial policy is the issue of why Indian electronics policy opted for entry, selectivity, and phasing policies of blanket ISI stressing industrial equipment and neglected international competitiveness. We have explained this as being due mainly to the exigencies of politics at the time of setting up the electronics policy institutions. It was a deliberately chosen overall policy mix meant to serve the political purposes of the top leadership. Economic concerns were subordinated to political priorities. The alternative liberalized ISI framework then competing with it was rejected for partisan political reasons and not for the lack of strategic capacity. While it may be true that the partial character of strategic capacity at the overall policy level did create distortions in the broad ISI regime that constrained and gave unique twists to electronics policy, deliberate political choices rather than strategic incapacity were the primary cause of the idiosyncrasies of Indian ISI and, in turn, the cause for its poor growth and export performance.

Our analysis of India's particular brand of ISI in its electronics policies over the past two decades goes beyond the issues of insulation of the state and its economic institutional consolidation to "high political" and policy flexibility factors. These consist of the macropolitical strategy of the state elite and the geopolitical/national security imperatives conditioning such strategy. These factors had a determining impact on Indian ISI, both at the outset and in the liberalizing 1980s. In fact, Indian strategy cannot be understood without them. The strategic capacity concept, with its elaborate checklist of defining characteristics, is misleading when taken out of context and applied in a mechanical way. It needs to be situated in the context of the "high politics" of the state elite's domestic and geopolitical grand strategy. These factors—especially the latter—tend to be neglected by post-dependency bargaining school scholars.[4] Strategic capacity, the possession of which makes strategy-led development possible, is quintessentially a concept representative of this optimistic school of thought inspired by the extraordinary industrial growth of the East Asian NICs. Intelligent state policy can break out of the underdevelopment and dependence trap, into industrial competitiveness. Certain conditions must be met in order to do this—strategic capacity, as defined, must exist.

Our comparison of Indian with Korean and Brazilian electronics policies suggests that while strategic capacity is a heuristically rich concept, it is inadequate. At best, it can rule out the possibility of states not possessing such a capacity being able to implement successfully a strategy-led development program in new industries. But it cannot quite explain or predict the specific policy trajectory or the success or failure of states apparently possessing such a capacity in sufficient measure.

We argue, therefore, that the concept of strategic capacity needs to be modified by introducing the following refinements:

(1) The concept of political closure does not adequately differentiate the variety of combinations of autonomy/insulation of the state from class interests. The kind of strategy chosen under different types of political closure can be very different and can have very different consequences for the development of the industry. Brazilian ISI differed from the Indian version in its policies on MNCs

versus domestic firms, and public versus private sectors, with weighty consequences for electronics strategy.

(2) The concept of economic institutional consolidation is inadequate by itself. It has to be conceived of in terms of levels of policy. In the Indian case, we have fruitfully distinguished between the overall industrial policy and electronics policy levels. This is especially important if one level of policy constrains the other(s), as was the case in Indian electronics.

(3) Strategic capacity concerns itself with relations between the state and the dominant social interest groups. In ignoring the relations between entities within the state's sphere of control, such as between public enterprises and policymaking institutions in India, it both adopts an undifferentiated view of the state and misses out an important dimension of insulation and, hence, strategic capacity.

(4) From the above points it follows that the concept of strategic capacity essentially deals with the state-as-machine and the relations between the state and its social–coalitional base—that is, state–society relations. It is necessary, as we have argued, to get beyond these levels to that of "high politics"—that is, of the grand strategies, domestic and geopolitical (including, very importantly, national security imperatives), of the state elite, especially of the top leadership. Unless the grand strategies of state elites dictate international competitiveness as a goal, this may not be aimed at even where an adequate degree of political closure and economic institutional consolidation does exist, as in India and in Brazil after 1979. The state elite's domestic and international grand strategy derives from strategic priorities, conditioned by the political–economic environment and the inherited ideological legacy.

(5) The state-as-machine and state–society relations levels of explanation also ignore the role of ideology at the level of policymakers and the top leadership.

(6) Policy flexibility is a crucial dimension of strategic capacity, since rapid adaptation to a changing techno–economic environment is vital to promoting competitiveness. It is constrained by *both* the interests and ideological legacy created by the policy path historically followed.

Our analysis of Indian electronics policy concludes, therefore, that the strategic capacity concept needs to be refined and augmented by the addition of strategic priorities and policy flexibility. These have to be in line with appropriate policies for industrial promotion in the international economic context. Therefore, when one looks at the historical, if not contemporary, possibility of much of the semi-industrialized Third World replicating the experience of successful NICs in creating competitive industries and being able to break out of dependency, one cannot start with state–society relations as the level of analysis, as the optimistic dependency and postdependency bargaining theorists do; one has to situate that level in the more macro level of the overall political strategies (including geopolitical) of state elites. Our extended concept of strategic capacity consists of these factors as key variables, grafted onto the concept as it emerges from the works of Deyo, Wade, Evans and Tigre, and Nayar.

NOTES

1. Johnson, "Political Institutions," p. 141.

2. Fernando Henrique Cardoso, "The Characterization of Authoritarian Regimes," in David Collier (ed.), *The New Authoritarianism in Latin America* (Princeton, NJ: Princeton University Press, 1979), p. 36.

3. Grieco, *Between Dependency and Autonomy*, p. 11.

4. Grieco, *Between Dependency and Autonomy*, is a good example. In his study of the Indian computer industry, he misses the national security determinants of Indian computer policy altogether, even while emphasizing the key role of the atomic energy policy group in the EC/DOE in the early years.

Selected Bibliography

Adler, Emmanuel. *The Power of Ideology: the Quest for Technological Autonomy in Argentina and Brazil*. Berkeley, CA: University of California Press, 1987.

____. "Ideological Guerrillas and the Quest for Technological Autonomy: Brazil's Domestic Computer Industry." *International Organization*, 40 (3), 1986.

Ahluwalia, Isher J. *Industrial Growth in India: Stagnation since the Mid-Sixties*. New Delhi: Oxford University Press, 1985.

Alam, Ghayur. "India's Technology Policy and its Influence on Technology Imports and Technology Development." *Economic and Political Weekly*, 20 (46/47), November 1985: 2073–2080.

Amsden, Alice. *Asia's Next Giant: South Korea and Late Industrialization*. New York: Oxford University Press, 1989.

Anchordoguy, Marie. *Computers, Inc.: Japan's Challenge to IBM*. Cambridge, MA: Harvard University Press, 1989.

Antonelli, Christiano. *The Diffusion of Advanced Telecommunications in Developing Countries*. Paris: OECD, 1991.

Baer, Werner, "Political Determinants of Development." In Robert Wesson (ed.), *Politics, Policies and Economic Development in Latin America*. Stanford, CA: Hoover Institution Press, 1984.

Balassa, Bela, et al. *The Structure of Protection in Developing Countries*. Washington, DC: International Bank for Reconstruction and Development, 1971.

Bardhan, Pranab. *The Political Economy of Development in India*. Oxford: Basil Blackwell, 1984.

Bastos, Maria Ines. "How International Sanctions Worked: Domestic and Foreign Political Constraints on the Brazilian Informatics Policy." *Journal of Development Studies*, 30 (2), January 1994.

Benn Electronic Publications. *Yearbook of World Electronics Data*, 1988–1992. Oxford: Elsevier Advanced Technology, various years.

Bhagwati, Jagdish N. *Foreign Trade Regimes and Economic Development: Anatomy and Consequences of Exchange Control Regimes*. New York: Columbia University Press, 1976.

____ and Richard A. Brecher. "National Welfare in an Open Economy in the Presence of Foreign-Owned Factors of Production." *Journal of International Economics*, X (1980).

Bharat Electronics Limited (BEL). Annual Reports.

Bharat Heavy Electricals Limited (BHEL). Annual Reports.

Bloom, Martin. *Technological Change in the Korean Electronics Industry*. Paris: OECD, 1992.

Bowonder, B., and K. Vinod Reddy. "Microelectronics: State of the Art." *Electronics: Information and Planning*, 19 (10), 1992.

Bowonder, B., and T. Monish Singh. "Information Technology: State of the Art." *Electronics: Information and Planning*, 19 (8), 1992.

Bowonder, B., and D. Poorna Chander Rao. "Microelectronics: State of the Art and Imperatives for India." *Electronics: Information and Planning*, 20 (8), 1993.

Brazil: Secretaria Especial do Informatica (SEI), *Boletim Informativo*, August 1987, Ed. Especial.

Brecher, Richard A., and Carlos F. Diaz-Alejandro. "Tariffs, Foreign Capital and Immiserising Growth." *Journal of International Economics*, 7 (1977).

Brundenius, Claes, and Bo Goransson. "Technology Policies in Developing Countries: The Case of Telecommunications in Brazil and India." *Viertelsjahresberichte*, 103, March 1986.

Brunner, Hans-Peter. "Building Technological Capacity: A Case Study of the Computer Industry in India." *World Development*, 19 (12), 1991.

Buffie, Edward F. "Financial Repression, the New Structuralists, and Stabilization Policy in Semi-Industrialized Economies." *Journal of Development Economics*, 14 (1984).

Business Korea, various issues.

Byun, Byung Moon. "Growth and Recent Development of the Korean Semiconductor Industry." *Asian Survey*, 34 (8), August 1994.

Bureau of Industrial Costs and Prices (BICP), Ministry of Industry, India. "Report on Electronics." New Delhi: BICP, December 1987.

____. "Report on Computer and Peripherals," Volume 1. New Delhi: BICP, January 1989.

Cardoso, Fernando Henrique. "The Characterization of Authoritarian Regimes." In David Collier (ed.), *The New Authoritarianism in Latin America*. Princeton, NJ: Princeton University Press, 1979.

Chudnovsky, Daniel, Masafumi Nagao, and Staffan Jacobsson. *Capital Goods Production in the Third World: An Economic Study of Technology Acquisition*. New York: St. Martin's Press, 1983.

Choue, In-won. "The Politics of Industrial Restructuring: South Korea's Turn Toward Export-Led Heavy and Chemical Industrialization 1961–74." Ph.D. dissertation, Department of Political Science, University of Pennsylvania, 1988.

Chun, Sang-Yil. "Direct Foreign Investment in Korea." Korea Exchange Bank, *Monthly Review*, 14 (11), November 1980.

Chung, Jee-man. "The Electronics Industry in Korea." Korea Exchange Bank, *Monthly Review*, 17 (10), November 1983.

Cole, David C., and Princeton N. Lyman. *Korean Development: the Interplay of Politics and Economics*. Cambridge, MA: Harvard University Press, 1971.

Dataquest (New Delhi), various issues.

Department of Electronics (DOE), India. Annual Reports.

____. *Electronics Information and Planning (EIP)*, New Delhi: Department of Electronics (DOE), Government of India, numerous articles and reports in various issues.

____. "Report of the Working Group on Electronics Industry for the Eighth Five-Year Plan." *Electronics: Information and Planning*. 18 (2), November 1990, and 18 (3), December 1990.

____. "Report of the Study Team on Science and Technology for the Eighth Five-Year Plan for the Electronics Industry." *Electronics: Information and Planning*. 17 (10–11), July–August 1990.

____. "Compendium of Electronics Policy and Procedures," Appendix 1–7, *Electronics: Information and Planning*, 15 (1), October 1987, for the full text of the following policy statements between 1981 and 1986: (1) "Policy on Electronic Components (1981)"; (2) "Industrial and Licensing Policy for Colour Television Receiver Sets (25 February 1983)"; (3) "Measures to Further Accelerate the Rapid Development of Electronics (18 August 1983)"; (4) "Manufacture of Telecommunication Equipment Relaxation from 100% Public-Sector Manufacture (23 March 1984)"; (5) "New Computer Policy (19 November 1984)"; (6) "Integrated Policy Measures in Electronics (21 March 1985)"; (7) "Policy on Computer Software Export, Software Development and Training (19 December 1986)."

____ and Indian Posts and Telegraphs (P & T) Department. "Report of the Inter-Departmental Working Group on Electronic Switching Systems Policy." New Delhi: DOE and Indian P & T Department, July 1979.

____. "Report of the Semiconductor Committee." *Electronics: Information and Planning*, 1 (2), November 1973 and 1 (3), December 1973.

____. "Report of the Panel on Minicomputers." *Electronics: Information and Planning*, 1 (5), February 1974.

____. "Ten-Year Perspective Plan for Electronics and Communications." New Delhi: DOE, 1975.

____. "Report of the Review Committee on Electronics" (Sondhi Committee). *Electronics: Information and Planning*, 7 (6 and 7), March and April 1980.

Department of Informatics and Automation Policy (DEPIN), Secretariat for Science and Technology, President of the Republic, Government of Brazil. *Panorama do Setor de Informatica*. Brasilia: DEPIN, September 1991.

Department of Telecommunications, India. Annual Reports, various years (see also Ministry of Communications, India).

Desai, Ashok V. "Indigenous and Foreign Determinants of Technological Change in Indian Industry." *Economic and Political Weekly*, 20 (46/47), November 1985.

Deyo, Frederic C. "Coalitions, Institutions, and Linkage Sequencing: Toward a Strategic Capacity Model of East Asian Development." In Deyo (ed.), *The Political Economy of the New Asian Industrialism*. Ithaca, NY: Cornell University Press, 1987.

Dhar, Ranajit, Gautam Biswas, Vijay Kumar Gupta, and A. V. Rama Rao. "Market Potential of Indian Electronics Industry During the Seventh Plan 1985–86 to 1989–90." New Delhi: DOE, 1985.

Dorfman, Nancy S. *Innovation and Market Structure: Lessons from the Computer and Semiconductor Industries*. Cambridge, MA: Ballinger, 1978.

Economic Times, New Delhi, 15 May 1985.

Edquist, Charles, and Staffan Jacobsson. "The Integrated Circuit Industries of India and the Republic of Korea in an International Techno–Economic Context." *Industry and Development*, 21, March 1987.

ELCINA. *ELCINA Directory 1984–85.*

Electronic Business, 1 June 1984, and various issues.

Electronics, various issues.

Electronics Corporation of India (ECIL). Annual Reports.

Electronic Industries Association of Korea (EIAK). *Statistics of Electronic and Electrical Industries: Production, Export and Import*. Seoul: EIAK, various years.

Erber, Fabio Stefano. "The Development of the Electronics Complex and Government Policies in Brazil." *World Development*, 13 (3), 1985.

Ernst, Dieter, and David O'Connor. *Competing in the Electronics Industry: the Experience of Newly Industrializing Economies*. Paris: OECD, 1992.

Evans, Peter B. "Class, State and Dependence in East Asia: Lessons for Latin Americanists." In Deyo (ed.), *The Political Economy of the New Asian Industrialism*. Ithaca, NY: Cornell University Press, 1987.

____. *Dependent Development: The Alliance of Multinational, State and Local Capital in Brazil*. Princeton, NJ: Princeton University Press, 1979.

____. "State, Capital and the Transformation of Dependence: The Brazilian Computer Case." *World Development*, 14 (7), 1986.

____. "Indian Informatics in the 1980s: the Changing Character of State Involvement." *World Development*, 20 (1), 1992.

____ and Paulo Bastos Tigre. "Going Beyond Clones in Brazil and Korea: a Comparative Analysis of NIC Strategies in the Computer Industry." *World Development*, 17 (11), 1989.

____ and Paulo Bastos Tigre. "Paths to Participation in 'Hi-Tech' Industry: a Comparative Analysis of Computers in Brazil and Korea." *Asian Perspective*, 13 (1), 1989.

Frankel, Francine R. *India's Political Economy: the Gradual Revolution 1947–77*. New Delhi: Oxford University Press, 1978.

Frischtak, Claudio. "The Information Sector in Brazil: Policies, Institutions and the Performance of the Computer Industry." Washington, DC: World Bank, Industrial Strategy and Policy Division, March 1986.

____. "Specialization, Technical Change and Competitiveness in the Brazilian Electronics Industry." Technical Paper No. 27. Paris: OECD Development Centre, October 1990.

Grieco, Joseph M. *Between Dependency and Autonomy: India's Experience with the International Computer Industry*. Berkeley and Los Angeles, CA: University of California Press, 1984.

Haggard, Stephan, and Chung-In Moon. "The South Korean State in the International Economy: Liberal, Dependent or Mercantile." In John Gerard Ruggie (ed.), *The Antinomies of Underdevelopment: National Welfare and the International Division of Labor*. New York: Columbia University Press, 1983.

Haggard, Stephan, and Tun-jen Cheng. "State and Foreign Capital in East Asian NICs." In Frederic C. Deyo (ed.), *The Political Economy of the New Asian Industrialism*. Ithaca, NY: Cornell University Press, 1987.

Hobday, Michael. "Telecommunications: A 'Leading Edge' in the Accumulation of Digital Technology? Evidence from the Case of Brazil." *Viertelsjahresberichte*, 103, March 1986, pp. 65–76.

____. "The Brazilian Telecommunication Industry: Accumulation of Microelectronic Technology in the Manufacturing and Service Sectors." Vienna: UNIDO/IS.511, 25 January 1985.

____. *Telecommunications in Developing Countries: The Challenge from Brazil*. London and New York: Routledge, 1990.

Indian Telephone Industries (ITI). Annual Reports.

International Monetary Fund. *International Financial Statistics 1995*. Washington, DC: International Monetary Fund, 1996.

International Herald Tribune, Paris, 2 June 1985.

Johnson, Chalmers. "Political Institutions and Economic Performance: The Government–Business Relationship in Japan, South Korea and Taiwan." In Frederic C. Deyo (ed.), *The Political Economy of the New Asian Industrialism*. Ithaca, NY: Cornell University Press, 1987.

Jones, Leroy C., and Il Sakong. *Government, Business and Entrepreneurship in Economic Development: the Korean Case*. Cambridge, MA: Harvard University Press, 1980.

Kim, Cae-One, Young Kon Kim, and Chang-Bun Yoon. "Korean Telecommunications Development: Achievements and Cautionary Lessons." *World Development*, 20 (12), 1992.

Kim, Hee-nan. "The Electronics Industry in Korea." Korea Exchange Bank, *Monthly Review*. 20 (5), May 1986.

Kim, Jung-hyun. "Recent Developments in R&D in Korea." Korea Exchange Bank, *Monthly Review*, 19 (12), December 1982.

Korea Development Bank. *Industry in Korea 1980*. Seoul: Korea Development Bank, 1980.

Krueger, Anne O. *Foreign Trade Regimes and Economic Development: Liberalization Attempts and Consequences*. Cambridge, MA: Ballinger, 1978.

____. "The Political Economy of the Rent Seeking Society." *American Economic Review*, 63 (3), June 1974.

Krugman, Paul R., and Lance Taylor. "Contractionary Effects of Devaluation." *Journal of International Economics*, 8 (1978).

Lall, Sanjaya. "India." *World Development*, 12 (5/6), May/June 1984.

Little, I. M. D., T. Scitovsky, and M. Scott. *Industry and Trade in Some Developing Countries: A Comparative Study*. London: Oxford University Press, 1970.

Malerba, Franco. *The Semiconductor Industry: the Economics of Rapid Growth and Decline*. Madison, WI: University of Wisconsin Press, 1985.

Mani, Sunil. *Foreign Technology in Public Enterprises*. New Delhi: Oxford & IBH, 1992.

Meemamsi, G. B. *The C-DOT Story*. New Delhi: Kedar Publications, 1994.

Ministry of Commerce, India. "Report of the Committee on Electronics Exports" (Menon Committee). *Electronics: Information and Planning*, 6 (2), November 1978.

Ministry of Communications, India. "Report of the Committee on Telecommunications" (Sarin Committee). New Delhi: Ministry of Communications, 1981.

Mody, Ashoka. "Institutions and Dynamic Comparative Advantage." *Cambridge Journal of Economics*, 14, 1990, 291–314.

____. "Planning for Electronics Development." *Economic and Political Weekly*, 22 (26), June 27, 1987.

Morehouse, Ward, and Ravi Chopra. *Chicken and Egg: Electronics and Social Change in India*. Lund: Research Policy Institute, Occasional report series no. 10, 1983.

Mowery, David C. "Innovation, Market Structure and Government Policy in the American Semiconductor Electronics Industry: A Survey." *Research Policy*, 12 (4), August 1983.

Munck, Ronaldo. "State Intervention in Brazil: Issues and Debates." *Latin American Perspectives*, 6 (4), Fall 1979.

Nayar, Baldev Raj. "The Politics of Economic Restructuring in India: the Paradox of State Strength and Policy Weakness." *Journal of Commonwealth and Comparative Politics*, 30 (2), 1992.

Nayyar, Deepak. "Transnational Corporations and Manufactured Exports from Poor Countries." *Economic Journal*, 88, March 1978.

Nilekani, Rohini. "Plugging the Leaks." *Imprint,* Bombay, November 1984, p. 11.

Ninan, T. N. "Restrictive Practice." *India Today*, New Delhi, 31 August 1984, pp. 100–101.

North, Douglass C. *Structure and Change in Economic History*. New York: W. W. Norton, 1981.

Oberoi, S. S. "Software Technology Park: Concept, Status and Procedure." *Electronics: Information and Planning*. 19 (3), December 1991.

O'Donnell, Guillermo. "Reflections on the Patterns of Change in the Bureaucratic–Authoritarian State." *Latin American Research Review*, 8 (1), 1978.

Olson, Mancur. *The Rise and Decline of Nations: Economic Growth, Stagflation and Social Rigidities*. New Haven, CT: Yale University Press, 1982.

Pack, Howard, and Larry E. Westphal. "Industrial Strategy and Technological Change." *Journal of Development Economics*, 22 (1986).

Parthasarathi, Ashok. "Informatics for Development: the Indian Experience." Paper presented at the second session of the North–South Roundtable of the Society for International Development, Tokyo, October 1987.

____. "Electronics in Developing Countries: Issues in Transfer and Development of Technology." Geneva: UNCTAD, TD/B/C.6/34, 1978.

PEICO (formerly Philips India), Annual Reports.

Piragibe, Clelia. "Policies Towards the Electronics Complex in Brazil." CNPq:

Centro de Estudos em Politica Cientifica e Tecnologica (CPCT), Texto No. 28, 1987.

Schmitz, Hubert. "Industrialization Strategies in Less Developed Countries: Some Lessons of Historical Experience." *Journal of Development Studies*, 21 (1), October 1984.

Scott-Kemmis, Don, and Martin Bell. "Technological Dynamism and Technological Content of Collaboration: Are Indian Firms Missing Opportunities?" *Economic and Political Weekly*, 20 (46/47), November 1985.

Sen, Pronab. "Software Exports from India: A Systemic Analysis." *Electronics: Information and Planning*, 22 (2), November 1994.

____. "National Telecom Policy: An Analysis." *Electronics: Information and Planning*, 22 (1), October 1994.

Sercovich, Francisco Colman. "Brazil." *World Development*, 12 (5/6), 1984.

Serra, Jose. "Three Mistaken Theses Regarding the Connection between Industrialization and Authoritarian regimes." In David Collier, ed. *The New Authoritarianism in Latin America*. Princeton, NJ: Princeton University Press, 1979.

Seshagiri, N. "Econometric Study: Economical Scale of Production for Television Receiver Industry under Indian Conditions." *Electronics: Information and Planning*, 1 (1), October 1973.

Special Secretariat for Informatics (SEI), Government of Brazil. *Panorama do Setor de Informatica*, 7 (16), Special Edition. Brasilia: SEI, August 1987.

Sridharan, E. "The Political Economy of Industrial Strategy for Competitiveness in the Third World: The Electronics Industry in Korea, Brazil and India." Ph.D. dissertation, Department of Political Science, University of Pennsylvania, 1989.

Subramanian, C. R. *India and the Computer: A Study of Planned Development*. New Delhi: Oxford University Press, 1992.

Telegraph, Calcutta, 14 May 1983.

Tigre, Paulo Bastos. *Technology and Competition in the Brazilian Computer Industry*. New York: St. Martin's Press, 1983.

Trebat, Thomas J. *Brazil's State-Owned Enterprises: A Case Study of the State as Entrepreneur*. New York: Cambridge University Press, 1983.

Tuomi, Helena, and Raimo Vayrynen. *Transnational Corporations, Armaments and Development*. New York: St. Martin's Press, 1984.

United Nations. *Yearbook of Industrial Statistics*. New York: United Nations, various years.

United Nations Center for Transnational Corporations. *Transnational Corporations in the International Semiconductor Industry*. New York: United Nations, 1983.

____. *Transborder Data Flows and Brazil*. New York: United Nations, 1983.

Varshney, Ashutosh. "The Political Economy of Slow Industrial Growth in India." *Economic and Political Weekly*, 19 (35), 1 September 1984.

Verma, R. K. "Electronic Hardware Technology Park: A Step Towards Globalisation." *Electronics: Information and Planning*, 20 (6), March 1993.

Vittal, N. "My Years with the Department of Electronics: Management of a Technical Department in Government." *Electronics: Information and Planning*, 20 (11), August 1993.

Wade, Robert. *Governing the Market: Economic Theory and the Role of Industrialization in East Asian Industrialization*. Princeton, NJ: Princeton University Press, 1990.

Wajnberg, Salomao. "The Brazilian Microelectronics Industry and its Relationship with the Communications Industry." Vienna: UNIDO/IS.546, 5 August 1985.

Westphal, Larry E. "The Republic of Korea's Experience with Export-Led Industrial Development." *World Development*, 6 (3), March 1978.

Westphal, Larry E., Yung W. Rhee, Linsu Kim, and Alice Amsden. "Republic of Korea." *World Development*, 12 (5/6), May/June 1984.

Wilson, Robert K., Peter K. Ashton, and Thomas P. Egan. *Innovation, Competition and Government Policy in the Semiconductor Industry*. Lexington, MA: Lexington Books, 1980.

World Bank. "India—Development of the Electronics Industry—A Sector Report." Washington, DC: World Bank, Report Number 6781-IN, Volumes I, II, and III.

____. "Korea: Development in a Global Context," 1984.

____. *World Development Report*. Washington, DC: World Bank, various years.

____. *Trends in Developing Economies 1992*. Washington, DC: World Bank, 1992.

____. *The East Asian Miracle: Public Policy and Economic Growth*. New York: Oxford University Press, 1993.

____. "Korea: Managing the Industrial Transition," Vols. 1 and 2. Washington, DC: World Bank, 1987.

Yoon, Chang-Ho. "International Competition and Market Penetration: A Model of the Growth Strategy of the Korean Semiconductor Industry." In Gerald K. Helleiner (ed.), *Trade Policy, Industrialization and Development*. Oxford: Clarendon Press, 1992.

Index

About the Author

ESWARAN SRIDHARAN is Associate Research Professor at the Centre for Policy Research in New Delhi. He has published numerous research papers and has held visiting fellowships at the Institute of Developing Economies, Tokyo, and London School of Economics.

ISBN 0-275-95418-8
90000>
EAN
9 780275 954185
HARDCOVER BAR CODE